George Sherbrooke Turpin

Lessons in organic chemistry

Part I: Elementary

George Sherbrooke Turpin

Lessons in organic chemistry
Part I: Elementary

ISBN/EAN: 9783337279813

Printed in Europe, USA, Canada, Australia, Japan

Cover: Foto ©berggeist007 / pixelio.de

More available books at **www.hansebooks.com**

ORGANIC CHEMISTRY

LESSONS

IN

ORGANIC CHEMISTRY

PART I.—ELEMENTARY

G. S. TURPIN, M.A. (Camb.), D.Sc. (Lond.)
PRINCIPAL OF THE TECHNICAL SCHOOL, HUDDERSFIELD

New York
MACMILLAN AND CO.
AND LONDON
1894

Norwood Press:
Berwick & Smith, Boston, U.S.A.

CONTENTS

ELEMENTARY

CHAP.		PAGE
1.	The Analysis of Organic Bodies	1
2.	Empirical and Molecular Formulæ	12
3.	Hydrocarbons of the Methane Series	19
4.	Olefines and Acetylene	28
5.	Haloid Derivatives	37
6.	The Alcohols	44
7.	Ethereal Salts—Ethers—Mercaptan	55
8.	Aldehydes and Ketones	61
9.	The Fatty Acids	68
10.	Acetyl Chloride and Acetic Anhydride	79
11.	The Amines	82
12.	The Amides and Amido-Acids	88
13.	Alkyl Compounds of Phosphorus, Arsenic, Silicon, and the Metals	92
14.	Glycol and its Derivatives. Succinic, Malic, and Tartaric Acids	99

CHAP.		PAGE
15. Lactic and Citric Acids	.	106
16. The Allyl Compounds	.	110
17. Glycerine and its Compounds	.	115
18. The Carbohydrates	.	119
19. Urea and Uric Acid	.	128
20. The Cyanogen Compounds	.	131

ELEMENTARY

CHAPTER I

THE ANALYSIS OF ORGANIC BODIES

Province of Organic Chemistry.—Up to the beginning of this century it was generally supposed that all chemical substances might be sharply divided into two classes, according as their formation was, or was not, possible without the aid of living organisms; those compounds which had been obtained only from some animal or plant were called organic bodies, and the action of the mysterious "vital force" was believed to be necessary for their formation. In 1828, however, the German chemist Wöhler prepared urea, a typical organic substance, from inorganic materials by a chemical reaction of very simple character, and thus broke down the separation which up to that time had been maintained between inorganic and organic substances; but, as a matter of convenience, we still retain the name organic chemistry for a department of the science which is concerned with the chemistry of the compounds of two elements, carbon and hydrogen, and their numerous derivatives. Amongst these are included nearly all the substances formed by the complicated chemical processes lying at the base of life, whether animal or vegetable, as well as a still larger number which have been prepared artificially by the simple processes of the laboratory. Many compounds obtained, in the first place, from animals or plants have been afterwards manufactured in the laboratory, and chemists have good reason to believe that in the future there will be no single substance known whose formation cannot be brought about by ordinary chemical reactions.

The distinction between organic and inorganic chemistry is, then, merely a convenient division of the vast material of the science, and organic chemistry may be defined as the chemistry of the hydrocarbons and their derivatives.

Reasons for the separate Study of Organic Chemistry. — The reasons which make it convenient still to maintain an artificial separation between inorganic and organic chemistry are chiefly the immense number of organic compounds known — a number which receives additions every day — and the different character of the problems presented to us by them as compared with the much less numerous and comparatively simple inorganic bodies. In organic chemistry the most important points claiming our attention are the grouping of the atoms present in each compound, and the influence of this grouping on the properties of that compound. We shall find cases where the molecules of two different substances contain exactly similar atoms, and in the same number, but the different arrangement of the atoms in the two molecules produces bodies with markedly different properties.

The name isomerism is given to this phenomenon. Cases of it are almost unknown in inorganic chemistry, but are extremely frequent in organic chemistry; see p. 24 for a further account of it.

On the other hand, we often find a series of organic compounds, all of different composition, but possessing very similar properties, owing to the presence in their molecules of the same group of atoms; prominent instances of this are furnished by the homologous series, of which more is said on p. 21.

Elements present in Organic Compounds. — Every organic compound contains carbon, and in nearly every one hydrogen is also found; the other elements may any of them occur, but those more frequently found are nitrogen, oxygen, the halogens, sulphur, and phosphorus.

The carbon, hydrogen, and nitrogen are most satisfactorily detected by heating the substance with copper oxide in a hard-glass tube; the organic material is burned up by the oxygen of the copper oxide, some of which is reduced to metallic copper, and the products of the combustion are carbon dioxide, water, and nitrogen gas. If no water is produced by the combustion, then no hydrogen was present in the substance examined, and if no nitrogen be given off, that element was similarly absent from the material employed.

EXPT. 1. Take a piece of ordinary combustion tubing closed at one end; introduce into it (*a*) enough *dry* cupric oxide (best granulated) to fill about three inches of the tube; (*b*) then about one gram of sugar; (*c*) and lastly, fill the tube nearly to the open end with more granulated copper oxide. Close the open end of the tube with a well-fitting rubber stopper, through which passes a piece of glass tubing carrying a small bulb, and connect this with two small wash-cylinders, of which the first contains lime-water (better, baryta water), and the second strong solution of caustic soda.

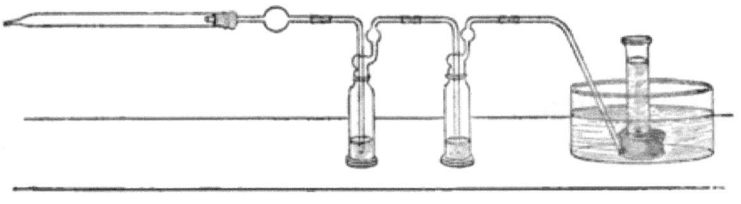

FIG. 1.—Apparatus for Experiments 1 and 2.

Heat the tube carefully in a combustion furnace, or over a row of four Bunsen burners; first applying heat at the two ends of the tube, and when these have become just red-hot, turning up gradually the two middle burners. Notice the production of water and CO_2, and that no nitrogen escapes from the second wash-cylinder after the air has been all driven out.

EXPT. 2. Repeat, using urea instead of sugar. Notice the large amount of gas evolved which is not absorbed by the caustic soda. Urea contains nearly fifty per cent of nitrogen.

When proper precautions are applied this method of combustion with copper oxide enables us to determine with considerable accuracy the amounts of carbon, hydrogen, and nitrogen present in any organic substance. The carbon and hydrogen are usually determined by one experiment, the nitrogen by a second.

Quantitative Determination of Carbon and Hydrogen.—There are several variations in the details of the experiment as adopted by different chemists; we shall describe only one plan of work.

A piece of hard-glass tubing, long enough to project about an inch from each end of the combustion furnace, is connected at the one end with a tube, through which either air or oxygen—in each case dry and free from carbon dioxide—may be supplied at will; and at the other end, with a U tube containing lumps of porous $CaCl_2$, to absorb and retain the water produced in the combustion, followed by a potash apparatus (Fig. 5), which similarly absorbs the carbon dioxide.

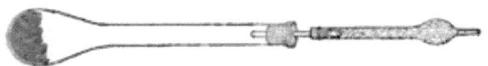

FIG. 2.—Flask fitted with $CaCl_2$ tube, in which the copper oxide is allowed to cool after being dried by heating to redness.

About two-thirds of the combustion tube at the end nearer the absorbing tubes is filled with granulated copper oxide, kept in position by two plugs of copper gauze; behind this is a "boat" of porcelain or platinum, into which about one-fifth

FIG. 3.—Tube arranged for combustion in a current of Oxygen; the $CaCl_2$ tube only is shown attached.

of a gram of the substance to be analysed has been accurately weighed; and then follows a longer plug of oxidised copper gauze, whose object is to prevent any backward diffusion of the products of the combustion.

The copper oxide having been previously thoroughly dried, the portions of the tube on either side of the boat are first raised to a dull red heat, and the actual combustion is then begun by carefully and gradually applying heat to the substance in the boat, while a very slow current of pure air is passed through the apparatus. Towards the end oxygen is introduced in place of air, with the object of burning up any carbonaceous residue that may have been left in or about the boat, and then air is again passed, in order to sweep along any CO_2 or oxygen that may be in the tube or absorbing apparatus. (Oxygen is heavier than air, hence the need of leaving the absorbing apparatus filled with air at the end, as at the beginning, of the combustion.)

The $CaCl_2$ tube and the potash apparatus were, of course, weighed before the combustion was commenced; they are weighed again after it is over, and the increase of weight

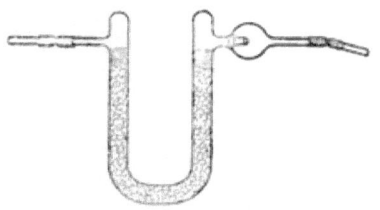

FIG. 4.—U-tube filled with $CaCl_2$ for absorbing the water produced in the combustion.

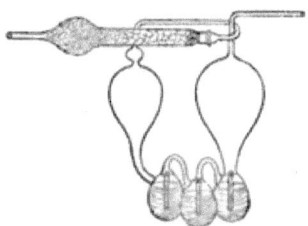

FIG. 5.—Potash apparatus for absorbing the CO_2.

gives the amount of water and of carbon dioxide produced by burning a known weight of the substance. An example will best illustrate how the percentage of carbon and of hydrogen present in the substance may be calculated.

Example. 0.2386 gram of a substance gave 0.4879 gram CO_2 and 0.0870 gram H_2O.

$$\text{The percentage of carbon} = 100 \times \frac{.4879}{.2386} \times \frac{3}{11} = 55.76,$$

for .4879 gram CO_2 contains $.4879 \times \frac{12}{44}$ gram of carbon.

$$\text{The percentage of hydrogen} = 100 \times \frac{.0870}{.2386} \times \frac{1}{9} = 4.05.$$

Quantitative Determination of Nitrogen. — There are several methods in use, but the only one which is applicable to all organic bodies alike is that of Dumas, in which the substance is burned with copper oxide in an atmosphere of carbon dioxide, and the liberated nitrogen collected over a solution of caustic potash and measured.

Fig. 6 represents the apparatus used. The combustion tube is filled with (*a*) a six-inch length of granulated oxide; (*b*) a mixture of a known weight of the substance with powdered copper oxide; (*c*) six or eight inches of granulated oxide; (*d*) a spiral of copper gauze. It is connected at one end

FIG. 6.

with an apparatus for evolving carbon dioxide, from which, first of all, a steady stream of the gas is passed for at least half an hour, until the air is entirely driven out from the tube. During this time the granulated oxide on either side of the mixture (*b*) may be cautiously heated. When the air is expelled the collecting apparatus is filled to the tap with

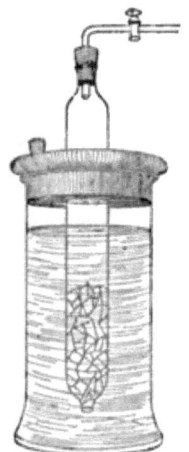

FIG. 7.—Apparatus for evolving a steady stream of CO_2 by the action of HCl upon marble.

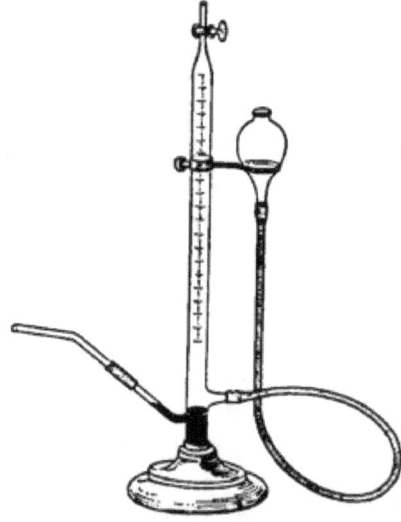

FIG. 8.—Apparatus in which the Nitrogen is collected over potash solution.

potash solution by raising the bulb, the tap is closed, and the stream of CO_2 stopped; the copper spiral is now heated to redness and the combustion proceeded with. At the end more CO_2 is passed, in order to sweep out any nitrogen from the tube and carry it into the measuring apparatus.

It is necessary to mix the substance with powdered CuO, as otherwise the combustion would not be complete in the atmosphere of CO_2. The object of the copper spiral is to reduce any oxides of nitrogen that might be evolved, and one must also be used in determining the carbon and hydrogen in a nitrogenous body.

An example will illustrate the method of calculation. The work is much simplified by the use of tables which have been specially prepared for the purpose.

Example. 0.2258 gram gave 28.3 c.c. moist nitrogen, measured at 9.5° C. and 765.5 mm.

The only difficulty is in calculating the exact weight of the nitrogen. It is measured over strong potash solution, whose vapour pressure at 9.5° C. is found in the tables as 7.1 mm.; the pressure of the nitrogen is therefore $765.5 - 7.1 = 758.4$ mm. Its volume (measured dry) at 0° and 760 mm. would therefore be

$$28.3 \times \frac{758.4}{760} \times \frac{273}{282.5} = 27.3 \text{ c.c.}$$

and its weight $27.3 \times .0000896 \times 14 = .03424$ gram (1 litre H weighs .0896 gram at normal temperature and pressure). The percentage of nitrogen is therefore $\frac{.03424 \times 100}{.2258} = 15.16.$

Many organic bodies containing nitrogen evolve ammonia when heated with soda lime (some, however, give off only part, and others none, of their nitrogen in the shape of ammonia), and on this plan it is possible in many cases to detect the presence of nitrogen, and estimate its amount. In the quantitative process (known by the names of Will and Varrentrapp) the liberated ammonia is absorbed by means of dilute hydrochloric acid placed in a bulb tube of suitable construction. The amount of the ammonia is ascertained by estimating how much hydrochloric acid has been neutralised by it.

This method has fallen into disuse, having been replaced by one due to Kjeldahl, which is applicable in all cases where Will and Varrentrapp's can be used, and is much more convenient. Kjeldahl decomposes the substance by heating it with concentrated sulphuric acid and addition of a little potassium permanganate. Under this treatment the nitrogen of the organic body is in many cases converted into ammonia, which is afterwards liberated by addition of caustic soda, distilled off and collected in a measured volume of dilute acid of standard strength. The calculation is precisely similar, whether Will's or Kjeldahl's method be adopted.

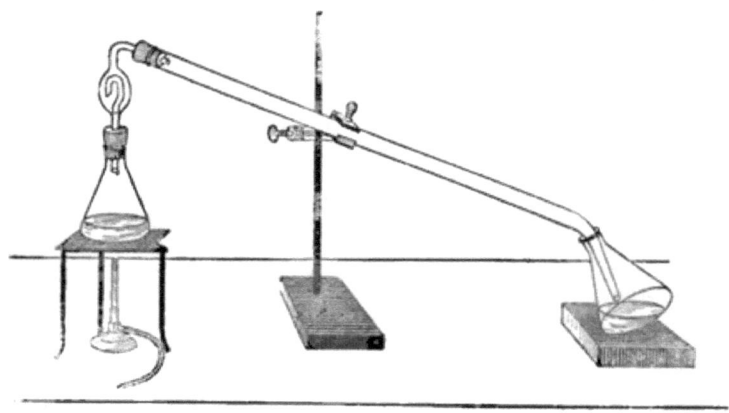

FIG. 9.—One form of apparatus for the second part of Kjeldahl's process; the ammonia is boiled off and absorbed by standard acid.

Example. 1.2350 gram of a substance was treated by Kjeldahl's method, and the ammonia produced collected in 25 c.c. of dilute hydrochloric acid of normal strength; at the end of the distillation it was found that 15.3 c.c. of normal soda solution were needed to neutralise the excess of acid which still remained uncombined.

The amount of ammonia produced was, therefore, sufficient to neutralise 9.7 c.c. ($=25-15.3$) of normal acid; that is to say, it was equal to the amount contained in 9.7 c.c. of a normal solution of ammonia. Such a solution contains 17 grams of NH_3 (molecular weight $=17$) in a litre, and in 9.7 c.c. there would be 17×9.7 milligrams NH_3; of this 14×9.7 mgms. are nitrogen, and therefore the percentage of nitrogen is $\dfrac{14 \times .0097}{1.2350} \times 100 = 11.0$.

Detection and Quantitative Estimation of the Halogens.—Organic substances containing chlorine, bromine, or iodine, do not, as a rule, react at all readily with silver nitrate; it is necessary first to decompose the organic matter, for which purpose either of the two following methods may be used:—

(*a*) Carius's method employs nitric acid as the oxidising agent. About .2 gram of the substance is introduced along with 1 or 2 c.c. of fuming nitric acid and a crystal of silver nitrate into a tube of stout glass ("pressure" tubing of fairly soft glass with walls 2 to 3 mm. thick is the most convenient) about 40 cm. long and 2 cm. external diameter. The open end of the tube is next carefully heated in the blow-pipe flame until the walls have thickened considerably at the heated spot, and then cautiously drawn out into a thick-walled capillary tube, which is finally sealed. The tube so prepared is heated in a specially designed and very strong air bath (or "cannon") to a temperature which varies, according to the character of the substance to be analysed, from 150° to 300° C. for one or two hours. The tube must be allowed to cool inside the "cannon," and even when cold contains gases (carbon dioxide and oxides of nitrogen) under such considerable pressure that its opening can only be safely effected by heating the capillary tip of the tube in a flame until the softened glass gives way before the internal pressure, and allows the compressed gases to escape.

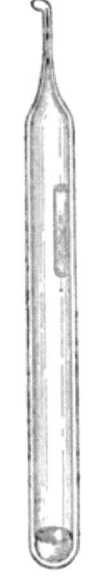

FIG. 10.—Sealed glass tube for Carius's method of analysis.

The silver chloride (or bromide or iodide) formed is washed out from the tube with distilled water, collected on a filter, washed, dried with the needful precautions by heating to fusion in a porcelain crucible, and weighed.

(*b*) The alternative or dry method consists in heating the substance with pure lime in a combustion tube heated in an ordinary combustion furnace. The calcium chloride (or bromide or iodide) produced is estimated in the usual way by precipitation with silver nitrate.

Example. (The calculation is precisely similar in both cases.) .1638 gram of the substance yielded .0953 gram AgCl.

The percentage of Cl is therefore, since 145.4 parts of AgCl contain 37.4 of chlorine,

$$\frac{.0053}{.1638} \times \frac{37.4}{145.4} \times 100$$
$$= 14.96.$$

The detection of the halogens can most certainly be accomplished by applying roughly one of the quantitative methods mentioned above; but more conveniently by Beilstein's plan, in which a little copper oxide, supported in a small loop at the end of a platinum wire, is heated in a Bunsen flame until this is no longer coloured, and is then used to convey a small portion of the substance adhering to the copper oxide into the flame. If chlorine is present copper chloride will be produced, and its vapour will give the characteristic blue and green flame of copper.

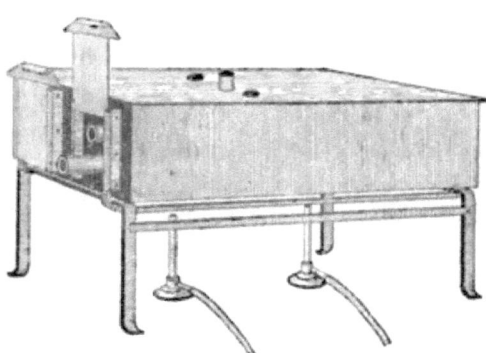

FIG. 11.—Air-bath for Carius's method.

Sulphur and **Phosphorus** may be estimated by heating the substance with fuming nitric acid in a sealed tube (Carius's method; see under Halogens). The sulphuric acid formed may be determined as barium sulphate, the phosphoric acid as magnesium pyrophosphate.

In the case of the less volatile substances, a dry method may conveniently be used, in which fusion in a silver dish with solid potassium hydrate, and gradual addition of potassium nitrate, is employed to effect the oxidation of the sulphur to sulphuric acid.

Example. .2178 gram of the substance gave .2586 gram of $BaSO_4$.

The percentage of sulphur is therefore, since 233 parts of $BaSO_4$ contain 32 parts of S,

$$\frac{.2586}{.2178} \times \frac{32}{233} \times 100$$
$$= 16.3.$$

The qualitative recognition of sulphur or phosphorus in an organic body may be effected by heating the dry substance with a little metallic sodium. If sulphur is present, sodium sulphide will be formed, and may be detected by the evolution of H_2S on addition of water and an acid, or by the use of sodium nitro-prusside, which gives an intense violet colouration with a trace of soluble sulphide. In the case of phosphorus, sodium phosphide (or if, as is advantageous, aluminium filings be employed, aluminium phosphide) is formed, from which the dampness of the breath is sufficient to evoke the characteristic smell of hydrogen phosphide.

Oxygen.—There is no convenient method known for the detection or estimation of oxygen in a compound. Its amount is determined by difference, *i.e.*, by subtracting the percentages of all the other elements present from 100, and taking the remainder to represent the percentage of oxygen.

Questions on Chapter I

1. Describe carefully the methods you would use for the quantitative estimation of the elements present in urea.
2. Explain how the percentage of nitrogen in an artificial manure can be readily determined.
3. Oil of mustard contains carbon, hydrogen, nitrogen, and sulphur. How would you prove that these elements and no others are present in it?
4. How is the percentage of chlorine in sodium chloride determined, and how must the method be modified in order to apply it to organic substances containing chlorine?

CHAPTER II

EMPIRICAL AND MOLECULAR FORMULÆ

THE **Empirical Formula** of a substance is the simplest formula which represents the results of analysis, and is calculated from these in the following way: Divide the percentage of each element by the corresponding atomic weight; find the smallest whole numbers standing in the same ratio as the quotients thus obtained, and you will have the indices of the formula. This is best illustrated by examples :—

A substance contains the percentages given below; to find its empirical formula

$$C = 40 \text{ per cent.}$$
$$H = 6.66 \text{ ,,}$$
$$O = 53.33 \text{ ,,}$$

Then

$$C = \frac{40}{12} = 3.33$$

$$H = \frac{6.66}{1} = 6.66$$

$$O = \frac{53.33}{16} = 3.33$$

and as these numbers are in the ratio 1 : 2 : 1, the empirical formula of the substance is CH_2O.

In the above example we have taken not the results of actual analysis, but the theoretical percentages. In calculating from the experimental numbers — always more or less inaccurate — we may sometimes have to choose between two or more formulæ which agree about equally well with the analytical results. In such cases it should be remembered that we usually find in a properly conducted analysis: (i.) about .1 or .2 per cent too little of carbon, unless halogens are present; (ii.) about .2 per cent in excess of hydrogen; and (iii.) about .2 or .3 per cent in excess of nitrogen (by Dumas's method).

The chief causes of these slight errors are: (i.) loss of CO_2 through incomplete absorption; (ii.) trace of moisture in the copper oxide employed; (iii.) presence of traces of air in the combustion tube and in the CO_2 used for expelling air from the tube.

The **Molecular Formula** represents not merely the results of analysis, but is also in agreement with whatever information we are able to obtain — by application of Avogadro's hypothesis or otherwise—as to the molecular weight of the compound. It is sometimes identical with the empirical formula, but is often a multiple of it, and the ratio is ascertained by a molecular weight determination. This may usually be effected by some one of the following methods.

1. **Chemical Methods** are not of very general application, and give only a *minimum* value of the molecular weight. Their principle is that in substituting one element (or radical) for another, we cannot replace a fraction of an atom. If, then, in a particular compound it is found possible to replace, say, one quarter of the hydrogen in it by some other element, without affecting the other three-fourths, we conclude that there were four atoms (or a multiple of four) in the molecule of that compound.

EXAMPLE I. The analysis of acetic acid leads to the empirical formula CH_2O; but there are numerous derivatives of the acid whose analysis shows that one-fourth only of the hydrogen has been replaced, such as monochloracetic acid $C_2H_3ClO_2$, silver acetate $C_2H_3AgO_2$, etc. Hence the molecular formula must contain four atoms of hydrogen, and is written $C_2H_4O_2$.

EXAMPLE II. Another substance, also possessing the same empirical formula CH_2O, is dextrose; but this compound yields a derivative in which analysis shows that five-twelfths of the hydrogen have been replaced, while seven-twelfths are left; there must then be not fewer than twelve atoms of hydrogen in the molecule, and its formula is put as $C_6H_{12}O_6$.

II. The **Physical Methods**—much more convenient, and in some respects more decisive, than the chemical—depend upon the "law of Avogadro," or upon its extension by Van't Hoff to the case of dilute solutions.

As applied to gases the law states that *at a given temperature the pressure of a gas is proportional to the number of molecules in unit volume.* If now we find the weights of equal volumes (at the same temperature and pressure) of two gases, we have the weights of equal numbers of molecules of the gases, and the ratio of these weights will give the ratio of the molecular weights. The *vapour density* of a substance is the ratio obtained by comparing the weight of a volume of that body (in the gaseous state) with the same volume of hydrogen at the same temperature and pressure. Admitting the weight of the hydrogen molecule to be 2, in accordance with the formula H_2, it follows that, when hydrogen is taken as the standard, *the molecular weight of any substance is twice its vapour density.* In determining this we do not need to measure both vapour and hydrogen under identical conditions, as by the help of Boyle's and Charles's laws we can easily reduce the results obtained to what they would be under the same pressure and temperature. The experimental processes which may be used for determining vapour densities are many, but the following are the most important :—

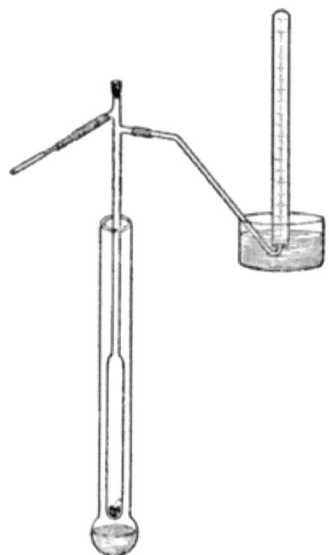

FIG. 12.—Apparatus for determining vapour-densities by V. Meyer's method.

(*a.*) **Victor Meyer's Method** has almost entirely supplanted the older ones of Hofmann and Dumas; the apparatus employed is shown in Fig. 12. A cylindrical bulb A, provided with a long and narrower neck, is heated to a steady temperature by some suitable means, usually by the vapour of a substance kept boiling in the jacketing tube. When the temperature has become quite steady, the cork is removed, and a small glass tube or thin bulb, containing about half a decigram of the body to be examined, is allowed to fall into the bulb, the cork being quickly replaced. In order that the experiment may succeed, it is necessary that the temperature of the bulb should be at least 20° or 30° C. above the boiling point of the compound under investigation, when this latter rapidly evaporates, and in doing so fills the lower part of the bulb with vapour, driving out through the side tube a corresponding volume of air, which is collected in E and measured.

It calculating the result it is unnecessary to know the temperature of A (it must, however, be steady). The vapour given off at the bottom of the bulb displaces its own volume of air, but this, before being measured, is cooled down to the temperature of the water over which it is collected. What we really obtain is, therefore, the volume which the vapour of the amount of substance used would occupy at the temperature and pressure in E. If, then, we divide the weight of substance used by the weight of the volume V of hydrogen (at the temperature and pressure in E), we have at once the vapour density of the compound examined.

The results, though not very accurate, are practically quite sufficient, as the question is usually not one of determining the exact molecular weight, but merely of the ratio of the molecular to the empirical formula.

Example. .0623 gram of alcohol gave by Victor Meyer's method 31.5 c.c. of air, measured at 15° C. and 750 mm. pressure.

This volume would become $31.5 \times \frac{273}{288} \times \frac{750}{760}$ c.c. at 0° C. and 760 mm.; and this volume of hydrogen (29.5 c.c.) would weigh $.0896 \times \frac{29.5}{1000}$ gram. The vapour density is therefore

$$\frac{.0623}{.0896} \times \frac{1000}{29.5} = \frac{\text{weight of substance}}{\text{weight of gas obtained reckoned as hydrogen}}$$

or 23.6.

(β.) Hofmann's Method is still occasionally made use of for substances which cannot readily be vapourised without decomposition under the ordinary pressure, though modifications of V. Meyer's method have also been made for this purpose. A long graduated tube, closed at one end, is filled with mercury, and inverted in a mercury trough, while round the upper portion of the tube a wider jacketing tube is placed, through which can be blown the vapour of some liquid of suitable boiling point. A small glass tube, containing about a fifth of a decigram of the substance, is introduced into the inner tube, and allowed to float up to the top of the mercury, where its contents are then vapourised on passing a current of steam or other vapour through the outside jacket.

On this method we require to notice the volume occupied by the vapour, and the height of the mercury in the inner tube above its level in the trough, besides knowing the temperature of the jacketing vapour. Its advantage is that the substance evaporates under a pressure considerably less than that of the atmosphere, in consequence of its partial compensation by the column of mercury in the inner tube.

Example. .0243 gram of substance vapourised at the temperature of boiling aniline ($183°$ C.) gave 54.5 c.c.; the height of the mercury column was 420 mm., that of the barometer 765 mm.

Now 54.5 c.c. of hydrogen at 345 mm. pressure (765 - 420) and $183°$ C. would weigh

$$54.5 \times \frac{.0896}{1000} \times \frac{345}{760} \times \frac{273}{456} = .00133 \text{ gram};$$

hence the vapour density of the substance is

$$\frac{.0243}{.00133} = 18.3.$$

Molecular Weight of Non-volatile Substances.—There are many substances of which it is quite impossible to determine the vapour density, as they are not volatile without decomposition. In such cases we can obtain assistance by applying methods, first established experimentally by Raoult, depending upon certain properties of solutions. Van't Hoff has brought forward a theory by which these various facts are connected together, but for the purpose in view the experimental data of Raoult are sufficient.

If we take 100 grams of water or any other solvent, and dissolve in it 1 gram of any substance, it is found that (*a*) the freezing point of the solution is lower, and (*b*) the boiling point is higher than that of the pure solvent. The amount of change is in each case dependent upon the molecular weight of the dissolved body. For the same solvent the change is proportional to the number of molecules dissolved in a given quantity of the solvent.

The apparatus devised by Beckmann for applying the first method, depending on the depression of the freezing point, is shown in Fig. 13. About 20 grams of the solvent are introduced into the central tube A of the apparatus, and the temperature being slowly brought down below the melting point, the exact temperature at which the solvent freezes is noticed on the thermometer. A small accurately weighed quantity of the substance to be examined is now introduced into the tube A, and made to dissolve by vigorous stirring and gentle warmth; then the temperature is again lowered, and when freezing occurs the freezing point of the solution is observed on the thermometer.

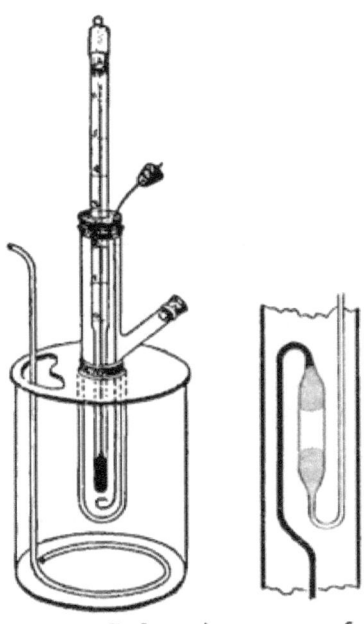

FIG. 13.—Beckmann's apparatus for determining molecular weights.

Let w = weight of substance used;
W =   ,,   solvent   ,,
t = difference between freezing point of the solution and freezing point of solvent;
m = molecular weight of the substance examined;

then the number of molecules of the substance dissolved in W grams of the solvent is $\dfrac{w}{m}$, and therefore in 100 grams of the solvent there would be, for a solution of the same strength,

$\frac{100\,w}{Wm}$ molecules dissolved. According to Raoult's results the depression of the freezing point for the solution is proportional to this number, and we have

$$t = K'\frac{100\,w}{Wm},$$

or
$$m = 100\,K'\frac{w}{Wt},$$

where K' is a constant depending on the nature of the solvent. The values of K' for the most important solvents are as follows:—

Water	19°
Benzene	49°
Naphthalene	74°
Acetic Acid	39°

For further particulars of this method, and of the similar one depending on the elevation of the boiling point, the student may advantageously consult *Outlines of General Chemistry*, by Ostwald, p. 137, or *Quantitative Analysis*, by Clowes and Coleman, p. 432.

QUESTIONS ON CHAPTER II

1. What reasons have we for writing the formula of acetic acid as $C_2H_4O_2$ instead of the simpler one CH_2O?

2. Describe Victor Meyer's method of determining vapour density. Calculate the vapour density of a substance from the following data: .0582 gram of the substance was used, and 23.5 c.c. of air were expelled (measured at 18° C. and 755 mm. pressure).

3. What methods can be used to determine the molecular weight of a substance such as sugar, which cannot be converted into vapour without decomposition?

4. Butyric acid has the empirical formula C_2H_4O, and silver butyrate is found to contain 55.4 per cent of silver; what do you conclude from these facts as to the molecular formula of the acid?

5. Calculate the molecular weight of a substance from the following results obtained by Raoult's method:—

Weight of acetic acid taken	20.5 grams
Freezing point of acetic acid	16.435° C.
Weight of substance dissolved	.1535 gram
Freezing point of solution	16.305° C.

CHAPTER III

HYDROCARBONS OF THE METHANE SERIES

Methane or marsh gas is theoretically the simplest of all the compounds of carbon and hydrogen. Analysis shows that its empirical formula is CH_4, and the fact that the gas is eight times heavier than hydrogen indicates the molecular weight sixteen, and shows that this simplest formula is also the molecular one.

It occurs naturally in the gas which occasionally comes off in bubbles from the bottom of stagnant ponds; in the "natural gas" escaping from fissures in the earth in certain oil-bearing districts, and constitutes the fire-damp of the coal miner; while ordinary coal-gas contains about one-third of its volume of methane.

Of methods used in the laboratory the three following are important, the first from the theoretical standpoint, and the two latter from that of practical work :—

1. Methane can be synthesised, *i.e.* built up from inorganic materials, by passing a mixture of H_2S with vapour of CS_2 over red-hot copper :

$$2H_2S + CS_2 + 8Cu = CH_4 + Cu_2S.$$

2. A convenient laboratory method, yielding, however, a somewhat impure methane, is to heat cautiously a mixture of sodium acetate with sodium hydrate (barium hydrate gives a less impure gas) :

$$\underset{\text{Sodium acetate.}}{NaC_2H_3O_2} + NaOH = Na_2CO_3 + \underset{\text{Methane.}}{CH_4}$$

If a glass vessel be used, it will soon be attacked by the melted caustic soda; and though this action can be lessened by using an admixture of quicklime (soda-lime is best), it is more convenient when possible to employ a copper retort.

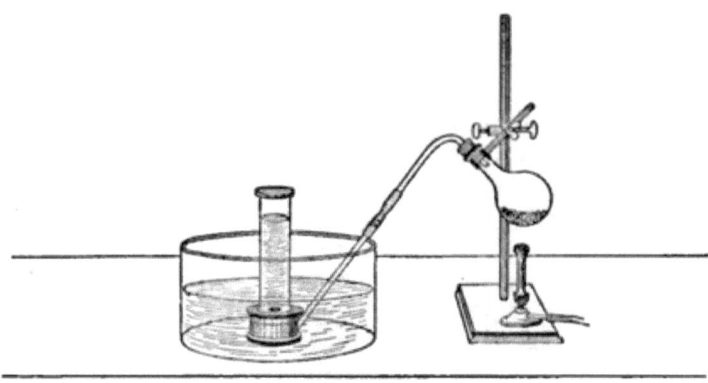

Fig. 14.—Apparatus for the preparation of CH_4 from sodium acetate and soda-lime.

EXPT. 3. Prepare marsh gas by heating some dry anhydrous (not crystallised) sodium acetate with about four parts of powdered soda-lime in a small glass flask fitted with cork and delivery tube. Collect two jars of the gas and examine its behaviour, (*a*) when a lighted taper is brought near, (*b*) when allowed to mix with bromine vapour contained in a second jar.

3. Pure methane is best prepared by the action on methyl iodide, CH_3I, of the zinc-copper couple in presence of alcohol. The couple is merely zinc covered with a deposit of copper by treatment with a solution of copper sulphate, and acts in presence of either water or alcohol as an excellent reducing agent:

$$CH_3I + H_2 = CH_4 + HI.$$
Methyl iodide.   Methane.

A more complete representation is given by the equation:

$$CH_3I + Zn + C_2H_6O = Zn \begin{cases} OC_2H_5 \\ I \end{cases} + CH_4.$$

Methane is a colourless gas, without taste or smell, only

slightly soluble in water, and very difficult to condense to a liquid. It burns in the air with a nearly non-luminous flame, which becomes much brighter if both the air and the methane are strongly heated before combustion (regenerative burners), and the products of the burning are water and carbon dioxide:

$$CH_4 + 2O_2 = CO_2 + 2H_2O.$$

A mixture of 1 vol. CH_4 with 2 vols. O_2 explodes violently when ignited. When strongly heated alone, methane is decomposed with formation of carbon, hydrogen, and smaller quantities of other products.

Methane is a very stable substance, and is not readily attacked even by the most active reagents. Nitric acid is almost without action upon it; chlorine and bromine attack it slowly (more quickly in sunlight than in the dark) with formation of "substitution products," in which one or more hydrogen atoms of the methane have been expelled (in combination with Cl or Br as HCl or HBr) and their place taken by halogen atoms:

$$CH_4 + Br_2 = CH_3Br + HBr,$$
Methyl bromide.

or
$$CH_4 + 2Br_2 = CH_2Br_2 + 2HBr, \text{ etc.}$$
Methylene bromide.

Homology.—Methane is the lowest member of a series of hydrocarbons, all of which can (in general) be prepared by similar reactions, and strongly resemble one another in their chemical behaviour. Each member differs from the one below it in the series by the replacement in its formula of an H atom by the group CH_3, to which the name methyl is given; the nett difference between any two successive members is therefore CH_2. Such a series is called a *homologous series*, and the study of organic chemistry is much simplified by the possibility of classifying in this way the immense number of known compounds into groups of similar bodies.

Starting from methane, CH_4, we have as the formula of the next member of the series $CH_4 + CH_2$ or C_2H_6; for the third C_3H_8, and so on up to $C_{60}H_{122}$, the highest which has yet been prepared. The generic formula is C_nH_{2n+2}.

Ethane, C_2H_6, stands next to methane, and can be prepared by similar reactions. In the first, we start not from sodium acetate (as for methane), but from the sodium salt of the acid next above acetic in the very important series of homologous acids, of which acetic forms the second and propionic the third member. Acetic acid is $C_2H_4O_2$ and propionic $C_3H_6O_2$. We proceed then as follows:—

1. Sodium propionate is heated with sodium hydrate,

$$NaC_3H_5O_2 \; + \; NaOH = Na_2CO_3 \; + \; C_2H_6$$
Sodium propionate.   Ethane.

when ethane is evolved and sodium carbonate remains.

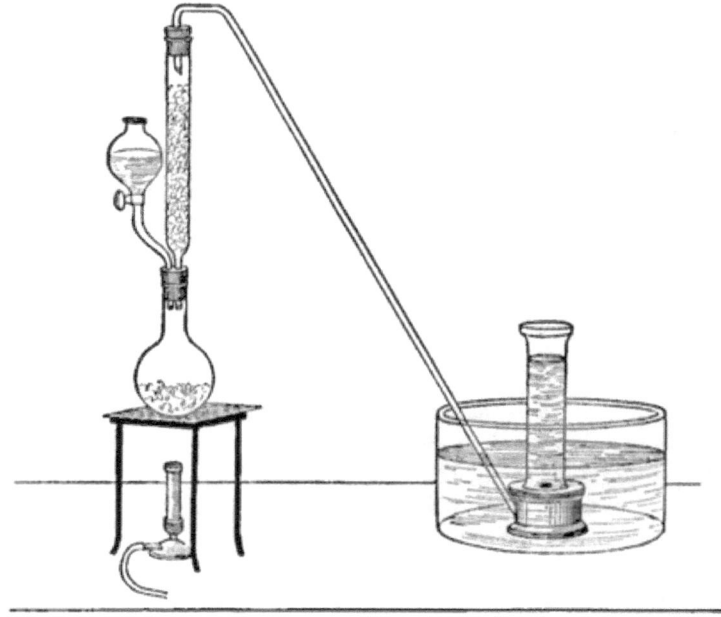

Fig. 15.—Apparatus for preparing C_2H_6 from ethyl iodide by the action of the zinc-copper couple.

2. In the second method for preparing ethane, ethyl iodide, C_2H_5I (homologous with methyl iodide, CH_3I), is reduced with the zinc-copper couple :

$$C_2H_5I + H_2 = C_2H_6 + HI.$$
Ethyl iodide. Ethane.

3. A third method is of a type applicable only to the preparation of those members of the series which contain an even number of carbon atoms in the molecule. In this case we start from methyl iodide, CH_3I, and by treating it with metallic sodium, abstract the iodine and cause two methyl residues—CH_3—to unite:

$$2CH_3I + 2Na = 2NaI + C_2H_6.$$
Methyl iodide. Ethane.

In accordance with this method of preparation, the formula of ethane may be written $CH_3 . CH_3$.

Ethane is a combustible gas, and burns with a more luminous flame than methane. It resembles that gas very greatly in chemical behaviour, and reacts in the same way with the halogens, forming substitution products, such as

$$C_2H_6 + Cl_2 = C_2H_5Cl + HCl.$$
Ethane. Ethyl chloride.

Propane, C_3H_8 ($= C_2H_6 + CH_2$), stands next above ethane. It may be prepared by methods corresponding to the first two of those given for ethane. The best is the following:

1. Propyl iodide, C_3H_7I ($= C_2H_5I + CH_2$), is reduced with the zinc-copper couple:

$$C_3H_7I + H_2 = C_3H_8 + HI.$$
Propyl iodide. Propane.

Butane is the name given to the next hydrocarbon of this series with the formula C_4H_{10}. We here encounter for the first time a fact of very great importance: that there may be, and often are, more substances than one corresponding to a particular molecular formula. This experimental fact we interpret to mean that two molecules, each containing the same atoms in the same number, may yet be distinct both chemically and physically; and this difference we explain as being due to the different arrangement of the atoms in the molecule. The name *isomerism* is given to this phenomenon, and substances which possess identical molecular composition, and yet differ from one another in the way described, are said to be *isomeric*. In the majority of such cases it is found possible to give a reasonable representation of the different chemical behaviour of the isomeric bodies by *structural formulæ*, which are also considered to represent more or less exactly the actual grouping of the atoms inside the molecule. Let us now consider more fully this particular case of the butanes.

In the first three members of this series no isomerism has been found to exist. Their formulæ, CH_4, C_2H_6, C_3H_8, may be expanded into CH_4, $CH_3.CH_3$, $CH_3.CH_2.CH_3$, which are in agreement with the valency hypothesis, and represent, more completely than the simple formulæ do, the modes of formation and general chemical behaviour of the three substances. In each case the formula is obtained from that of the next lower compound by substituting methyl, CH_3, for hydrogen. In methane there are four hydrogen atoms in the molecule, but these are all similarly circumstanced, and whichever of them be replaced we obtain the same ethane, $CH_3.CH_3$. Similarly, when we proceed from this to propane; the six hydrogen atoms in the ethane molecule are all of equal value, and we get always the same propane when any one of them is substituted by a methyl group. But at the next step this is no longer the case, for the eight hydrogen atoms in propane are not all similarly placed; while six of them are alike, and the other two also like one another in position, there is a difference between the atoms attached to the two terminal carbon atoms and those which are connected to the carbon atom in the centre of the chain. Hence two formulæ for butane may be deduced from that of propane by substituting CH_3 for H atoms of different value:

$CH_3.CH_2.CH_3$ gives (1) $CH_3.CH_2.CH_2.CH_3$
and (2) $CH_3.CH.(CH_3)_2$.

So far by paper work. Experimental investigation has proved that there are two butanes, each with the formula C_4H_{10}, and each rightly placed, according to its general behaviour, in the methane series.

Of these two butanes one is prepared from ethyl iodide by abstracting the iodine with sodium:

$$2CH_3.CH_2I + 2Na = 2NaI + CH_3.CH_2.CH_2.CH_3,$$
Ethyl iodide. Normal butane.

and this mode of formation is well represented by the formula given in the above equation. This particular butane is called *normal butane*, or simply butane. For the other butane the formula $CH_3.CH.(CH_3)_2$ remains, and the name given to it is *isobutane*.

Pentane, C_5H_{12}, is the generic name of the isomeric hydrocarbons corresponding to the formula given. If we attempt to work out the number of isomers which may in accordance with the valency theory be obtained, we find that *three* are possible; and experimental work has enabled us actually to prepare isomeric pentanes, and to assign to each, one of the three formulæ indicated by theory.

The most important is *normal pentane*, $CH_3.CH_2.CH_2.CH_2.CH_3$, which is contained in crude petroleum, and can be isolated from it as a volatile inflammable liquid boiling at 37°. This has been used as a means of obtaining a reliable standard of illumination for photometric purposes.

The following table illustrates the isomerism of the butanes and pentanes:—

Name.	Formula.	Preparation.
Normal Butane	$CH_3.CH_2.CH_2.CH_3$	From ethyl iodide and zinc dust: $2CH_3.CH_2I + Zn = C_4H_{10} + ZnI_2$
Isobutane	$CH_3.CH_2:(CH_3)_2$	From isobutyl iodide by reduction: $(CH_3)_2CH.CH_2I + H_2 = C_4H_{10} + 2HI$
Normal Pentane	$CH_3.CH_2.CH_2.CH_2.CH_3$	Separated from petroleum.
Isopentane (Dimethyl-ethyl-methane)	$CH_3.CH_2.CH:(CH_3)_2$	
Tetra-methyl-methane	$C(CH_3)_4$	

Petroleum and Paraffin.—In various parts of the world, more especially in Pennsylvania and in Baku, a province of Southern Russia, oil-bearing strata occur from which an inflammable oil can be obtained. Wells are drilled through the overlying layers of earth until the oil is struck at a depth varying from 50 to 2000 feet and over. In many cases the newly-opened well "spouts" oil, frequently with uncontrollable violence, but as the original gas-pressure declines, it becomes necessary to have recourse to pumping. The crude oil requires to be refined, and both in America and in Russia this process is carried on, not at the wells themselves, but at large refineries conveniently situated for export. The oil is transported to the refineries by means of long lines of pipes, through which it is forced by powerful pumps.

Investigation has shown that American and Russian petroleums differ essentially in chemical composition. American petroleum is almost entirely a mixture of various hydrocarbons of the methane series from CH_4 itself up to solid hydrocarbons of very high molecular weight. The refining of the crude oil has for its chief purpose the separation of this complex mixture into a number of fractions, and is accomplished by distillation. The more volatile portions are the first to come over, and are followed by others of higher and higher boiling points. The most important fractions are :—

(1) Gasoline, B.P. 30°-100°, used for making "oil-gas," which is simply air saturated with vapour of gasoline.

(2) Petroleum proper, B.P. 150°-300°, used in lamps.

(3) Higher boiling portions from which lubricating oils and vaseline are obtained.

Russian petroleum contains only a very small percentage of hydrocarbons of the methane series, the chief bulk being "naphthenes" of the generic formula C_nH_{2n}. These are distinct from the olefines of the same formula, and will not be considered until the second part of this book in connection with the benzene hydrocarbons, from which they are derived. The products obtained by refining the Russian crude oil are very similar to those from American petroleum, but a larger yield of oils suitable for lubricating machinery is got, and the residue is not usually worked up for a vaseline-like product, but is generally employed as fuel.

Another important source of hydrocarbons of the methane

series is the destructive distillation of bituminous shale or other material of similar composition. This process is largely carried on in the south-west of Scotland, and from the products various valuable mixtures of hydrocarbons are separated by refining. One of these is "paraffin oil"; another is the white solid "paraffin wax," and both are made up almost exclusively of hydrocarbons of the methane series.

Properties of the Hydrocarbons, C_nH_{2n+2}.—All the hydrocarbons of this homologous series, from marsh gas itself up to the highest member yet obtained, present an almost complete resemblance in chemical behaviour. They are all very inert substances, not attacked by nitric acid, and only gradually acted upon by chlorine or bromine. The products formed by the action of the halogens are substitution products, in which some of the hydrogen of the hydrocarbons has been replaced by chlorine or bromine. In no case are addition products formed by the members of this series.

The physical properties of the members change gradually as we pass from one end of the series to the other. The lowest members are gases requiring great pressure or cold to convert them into liquids; the pentanes are volatile liquids, and, ascending the series, we come to liquids of higher and higher boiling point; while still farther up the series we meet with hydrocarbons which are solid at the ordinary temperature.

Questions on Chapter III

1. Describe the preparation and properties of methane.
2. Methane and ethane are members of a homologous series; show the bearing of this statement upon the properties of the gas and the methods used for their preparation.
3. What is meant by isomerism? Deduce the formula of the two isomeric butanes from that of propane.
4. What substances are contained in crude petroleum? What commercial products are obtained from it, and by what processes?

CHAPTER IV

OLEFINES AND ACETYLENE

THE olefines form a second series of hydrocarbons, of which the starting-point is ethylene, C_2H_4. The succeeding homologues differ always by CH_2, and it thus follows that every member of the series has the same percentage composition. They differ, of course, in molecular weight, and therefore in vapour density. The generic formula is C_nH_{2n}, and while we again find the same similarity in general chemical behaviour between all the olefines, as between all the hydrocarbons of the C_nH_{2n+2} series, there are important differences between the two separate series. The chief of these are summed up in the contrast of the two terms *saturated* and *unsaturated*, applied respectively to the methane and to the olefine series.

Ethylene, C_2H_4, is the lowest known member of the series, and as in every reaction where we should expect a substance, CH_2, to be produced we obtain instead C_2H_4, it seems established that no compound of the formula CH_2 can exist. The most convenient way of preparing ethylene is by the action of concentrated sulphuric acid upon ethyl alcohol, C_2H_6O, a reaction which may very concisely be represented thus:

$$C_2H_6O - H_2O = C_2H_4.$$
Ethyl alcohol.   Ethylene.

What really happens is that ethyl-sulphuric acid, $C_2H_5 . HSO_4$, is first produced, and this, when heated, decomposes into ethylene and sulphuric acid;

$$C_2H_5OH + H_2SO_4 = C_2H_5 \cdot HSO_4 + H_2O$$
Ethyl alcohol.   Ethyl-sulphuric acid.

$$C_2H_5 \cdot HSO_4 = C_2H_4 + H_2SO_4.$$

EXPT. 4. Prepare ethylene by heating in a capacious flask (2 litres) a mixture of 40 c.c. methylated spirit with 200 c.c. concentrated H_2SO_4 along with some sand, to prevent frothing. The gas can be purified from SO_2 and other impurities by washing with solution of caustic soda, and can be collected over water.

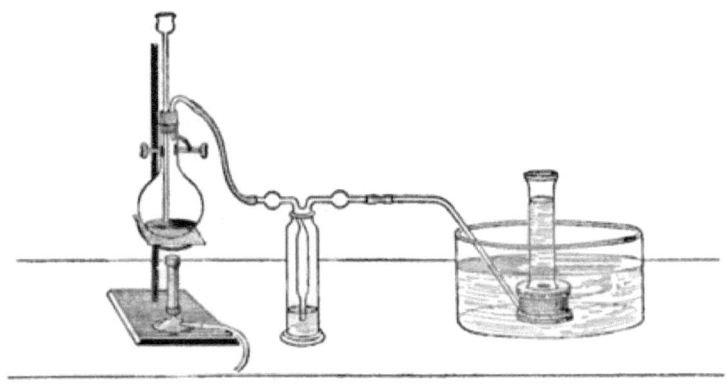

FIG. 16.—Preparation of Ethylene from alcohol and sulphuric acid.

Ethylene is a colourless gas with a faint sweetish smell, and burns in the air with a bright yellow flame. Like all the other members of the series, it is an *unsaturated* compound, being able to unite directly with several substances—Cl, Br, I, H, HI, etc., — nothing being driven out from the ethylene molecule to be replaced by the substituting atoms. This is explained on the valency hypothesis as due to those carbon valencies which are unsaturated in the molecule of ethylene.

Ethylene combines readily with Cl, Br, and I (in alcoholic solution) at the ordinary temperature, with hydrogen when the two mixed gases are passed over platinum black; ethane or a derivative of it is in each case the product. When passed into concentrated H_2SO_4, ethylene is absorbed and ethylsulphuric acid is formed :

$$C_2H_4 + H_2SO_4 = C_2H_5 \cdot HSO_4.$$

EXPT. 5. Fill two jars with ethylene and bromine vapour; bring them together mouth to mouth. The colour of the bromine is rapidly discharged, and small oily drops of a liquid—ethylene dibromide—are formed:

$$C_2H_4 + Br_2 = C_2H_4Br_2.$$

Try a similar experiment with methane and bromine. In this case the action takes place slowly, and the bromine makes its way into the methane molecule only by expelling some of the hydrogen:

$$CH_4 + Br_2 = CH_3Br + HBr.$$

Methane is termed a saturated compound, whereas ethylene is unsaturated.

Propylene, C_3H_6, is obtained most conveniently by a reaction typical of a second general method of preparing the olefines. Isopropyl iodide, C_3H_7I, a halogen derivative of the corresponding hydrocarbon of the methane series (in this case C_3H_8), is treated with an alcoholic solution of potash, whereby one atom of hydrogen and one of the halogen are abstracted:

$$\underset{\text{Isopropyl iodide.}}{C_3H_7I} + KOH = \underset{\text{Propylene.}}{C_3H_6} + KI + H_2O.$$

The formula of the isopropyl iodide is $CH_3 . CHI . CH_3$. The C_3H_6 obtained from this by abstraction of HI may be either $CH_3 . C . CH_3$ or $CH_3 . CH : CH_2$, but the first formula is negatived by numerous facts, of which a very conclusive one is that propylene furnishes three isomeric chloro-propylenes, C_3H_5Cl. This is readily explained by the second, but is quite inconsistent with the first. The two dots in the correct formula indicate that the two carbons, one on either side, are each capable of further uniting with an additional atom, and are connected together in a different manner from that prevailing between carbon atoms which are exerting their maximum valency. Two such carbon atoms as those we have been considering in the propylene molecule are said to be united by an *ethylene linkage*, or by a *double bond;* but it must not be imagined from the latter expression that such atoms are more firmly held together than those united in the ordinary way (single bond). As a matter of fact, an unsaturated molecule when it suffers decomposition usually breaks up most readily at the so-called double bond. Thermo-chemical investigation also shows that an ethylene linkage cannot be regarded as simply the double of single linkage ; there is a real difference in kind between those two modes in which carbon atoms may be connected.

Great assistance in correlating a very large number of experimental facts is furnished by the *tetrahedral theory of the carbon atom*, which was proposed by Van't Hoff in 1877, and has since been extended by Wislicenus and other chemists. According to this valuable hypothesis, the carbon atom is regarded as being similar in shape to a regular tetrahedron, a solid figure bounded by four equilateral triangles, and two carbon atoms may be connected together in the following three ways :

a. Simple linkage : the two tetrahedra are in contact at a corner of each.

b. Double linkage : the two tetrahedra are in contact along an edge of each.

c. Triple linkage : the two tetrahedra have a whole face of each in contact.

The first kind of linkage is exemplified in the case of ethane, the second in ethylene, and the third in acetylene. Substances which contain a double or triple linkage are unsaturated, and the theory affords a clear representation of the way in which an unsaturated body becomes saturated by addition of chlorine, bromine, etc. The following diagrams will illustrate these points better than any verbal explanations.

To return to propylene : the theory which we have been considering embodies very conveniently a large number of experimental generalisations, among them this, that such a formula as $CH_3 . C . CH_3$, in which carbon acts as a divalent element, represents an arrangement of atoms incapable of permanent existence. We have already seen reason to reject it in favour of the alternative $CH_3 . CH : CH_2$.

There are a few exceptional cases in which carbon is divalent, and where it is therefore necessary to suppose that two of the four corners of the carbon tetrahedron are unemployed, *e.g.* CO.

Butylene, C_4H_8, the next member of the series, furnishes an instance of isomerism. There are three isomeric butylenes, all of which may be looked upon as derived from ethylene by replacement of hydrogen by methyl or ethyl groups, and their names are best chosen to represent the manner of this derivation :

   a. Symmetrical dimethyl-ethylene, $CH_3 . CH : CH . CH_3$.
   b. Unsymmetrical dimethyl-ethylene, $CH_2 : C(CH_3)_2$.
   c. Ethyl-ethylene, $C_2H_5CH : CH_2$.

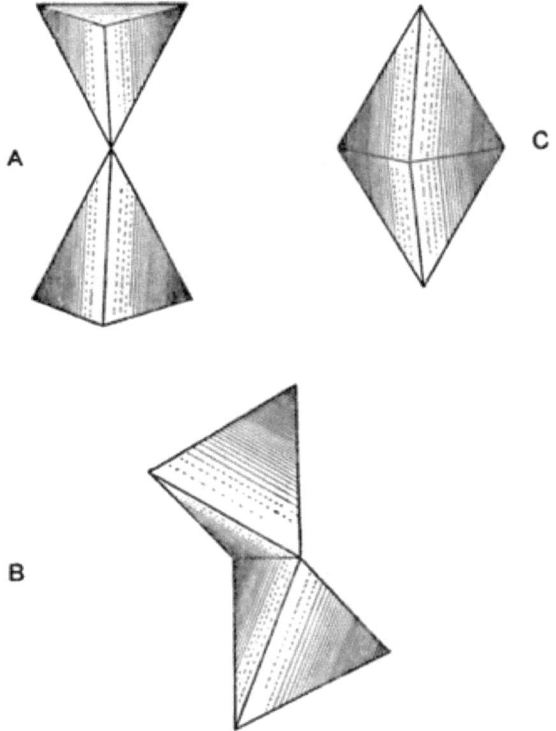

FIG. 17.—Representation of carbon atoms linked together as in (A) Ethane, (B) Ethylene, (C) Acetylene.

The names are somewhat cumbrous, but have the advantage of telling as much about the substances as the formulæ themselves; they are, in fact, merely the formulæ in words instead of symbols. The methods by which these three isomers have been prepared are too complicated for us to enter into; the important thing is that the theoretical number of isomers have been obtained.

Acetylene, C_2H_2, is the lowest member of another series of unsaturated hydrocarbons. In this we have a triple linkage between the carbons, so that only two hydrogen atoms can be attached to them, one to each : $HC:CH$.

Acetylene can be made by several different reactions :—

(1) By heating ethylene dibromide, $C_2H_4Br_2$, with an alcoholic solution of potash :

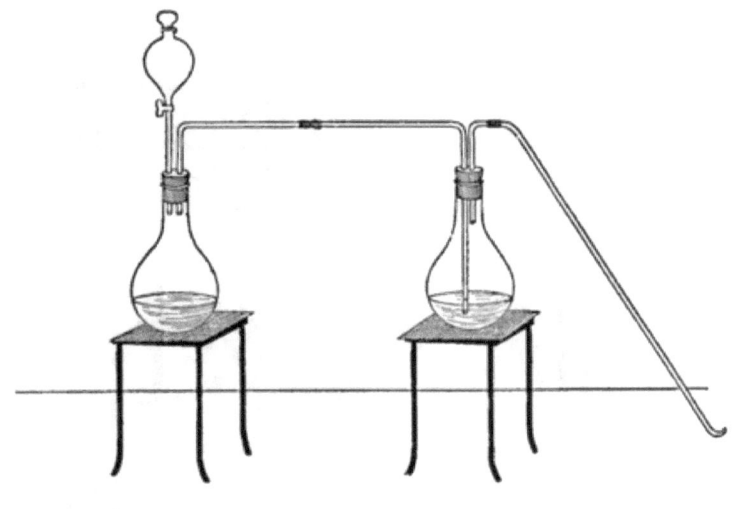

FIG. 18.—Preparation of Acetylene by the action of alcoholic potash on ethylene bromide; both flasks contain potash, and the bromide is allowed to drop slowly from the tap-funnel into the heated potash in the first flask.

$$CH_2Br . CH_2Br - 2HBr = HC:CH.$$
Ethylene bromide.   Acetylene.

(2) By direct combination of carbon and hydrogen, when the electric arc is passed between carbon poles in an atmosphere of hydrogen.

(3) By the action of water upon barium carbide :

$$BaC_2 + 2H.OH = C_2H_2 + Ba(OH)_2.$$
Barium carbide.   Acetylene.

This is a very convenient method of preparing the gas when it is wanted in considerable quantity.

(4) By the degraded combustion of coal-gas, such as occurs when the temperature of the flame is artificially lowered by contact with a metal surface (a Bunsen burner in which the gas is burning at the bottom of the brass tube):

$$C_2H_6 + O_2 = C_2H_2 + 2H_2O.$$
$$\text{Ethane.} \qquad \text{Acetylene.}$$

Acetylene is a colourless gas with an unpleasant smell. It is soluble in about its own volume of water, and burns in the air with a bright smoky flame. The most remarkable chemical characteristic of acetylene is its property of forming explosive compounds containing copper or silver. By means of these compounds acetylene can be readily detected and isolated from its mixture with other gases.

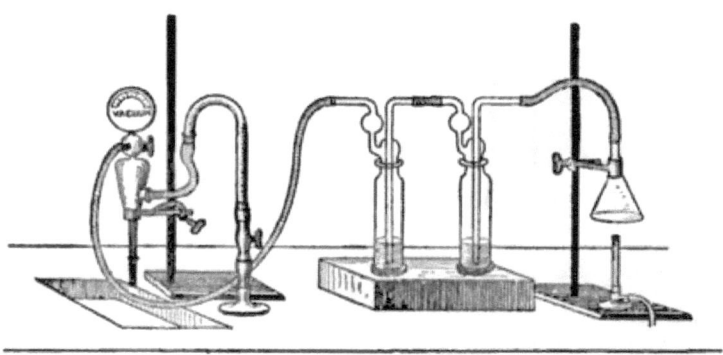

FIG. 19.—Preparation of Acetylene by the degraded combustion of coal-gas.

EXPT. 6. Prepare some cuprous chloride, CuCl, by passing SO_2 gas into a solution of 90 grams NaCl and 200 grams crystallised $CuSO_4$ until the gas is no longer absorbed; pour into about half a litre of water, and filter. The white precipitate of CuCl is collected, and dissolved in some strong ammonia solution.

Open wide the air-holes of a large Bunsen burner, and light the gas at the bottom of the tube. Over the burner support a funnel, which is connected with a gas-washing bottle containing the ammoniacal solution of cuprous chloride. Join the other tube of the wash-bottle to an aspirator, and draw a steady current of air through the apparatus.

A dark red precipitate will form in the solution of cuprous chloride. This has the composition C_2HCu, and when dry explodes on being heated, or if struck between two metal surfaces. Acetylene can be recovered from it by treatment with dilute hydrochloric acid.

Acetylene unites readily with hydrogen, when the two gases are passed over platinum black, to form ethane:

$$CH:CH + 2H_2 = CH_3 . CH_3.$$
Acetylene. Ethane.

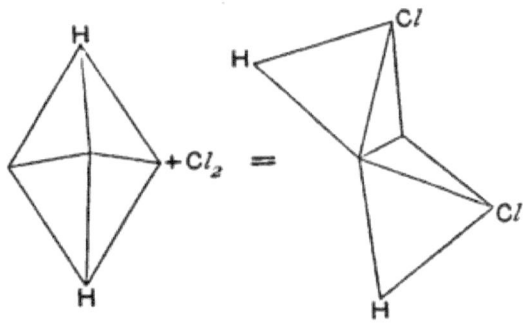

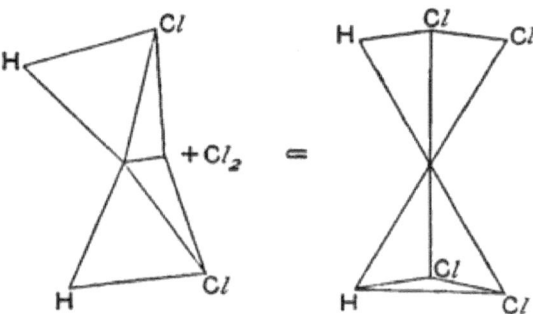

FIG. 20.—Representation on the tetrahedral theory of the conversion of Acetylene into dichlor-ethylene and tetrachlor-ethane by the action of chlorine.

Chlorine is without action upon the pure gas in the dark, but in sunlight dichlor-ethylene and tetrachlor-ethane are successively formed:

(a) $CH:CH + Cl_2 = CHCl:CHCl$,
   Acetylene.   Dichlor-ethylene.

(b) $CHCl:CHCl + Cl_2 = CHCl_2 . CHCl_2$.
   Dichlor-ethylene.   Tetrachlor-ethane.

These changes are represented on the tetrahedral theory by the diagrams in Fig. 20.

Questions on Chapter IV

1. What is the chief difference in chemical behaviour between a saturated and an unsaturated compound?

2. Give the preparation of acetylene and the most important properties of the gas.

3. Give an elementary account of the tetrahedral theory of Van't Hoff as applied to explain the structure of the three hydrocarbons C_2H_6, C_2H_4, C_2H_2.

4. How is ethylene made? What is formed when it is passed into concentrated sulphuric acid?

5. How could you prepare ethane from the elements carbon and hydrogen?

CHAPTER V

HALOID DERIVATIVES

In any hydrocarbon it is generally possible to replace from one up to the full number of hydrogen atoms present in the molecule, by chlorine, bromine, or iodine.

Methyl Chloride, CH_3Cl, is the first to be considered of all these haloid derivatives; it may be prepared

1. By the direct action of chlorine upon methane, according to the equation:

$$CH_4 + Cl_2 = CH_3Cl + HCl,$$

a process which is favoured by the influence of sunlight.

2. From methyl alcohol, CH_4O (a substance to be considered in the next chapter, when we shall learn that the formula is conveniently written $CH_3.OH$, in order to indicate the way in which the alcohol most readily reacts), by the action of various compounds containing chlorine. The equation is simplest in the case when HCl gas is used:

$$\underset{\text{Methyl alcohol.}}{CH_3.OH} + HCl = \underset{\text{Methyl chloride.}}{CH_3Cl} + H_2O,$$

a reaction which easily occurs when HCl is passed into boiling methyl alcohol, to which some zinc chloride (a very hygroscopic substance) has been added.

Methyl chloride is a gas with a pleasant smell, fairly soluble in water, and pretty easily condensed by cold or pressure to a liquid. It is used commercially in the manufacture of certain aniline dyes, and for this purpose is prepared from a by-product of the beet-sugar industry, and sold compressed in strong steel cylinders. It burns with a green flame.

Methene Chloride, CH_2Cl_2, can be obtained by the further action of chlorine upon methyl chloride:

$$CH_3Cl + Cl_2 = CH_2Cl_2 + HCl.$$

This is the second step in a series of successive substitutions of the hydrogen atoms by chloride; the third step yields

Chloroform, $CHCl_3$, which is, however, more readily prepared by a complicated reaction where ordinary ethyl alcohol, C_2H_6O, is treated with bleaching powder.

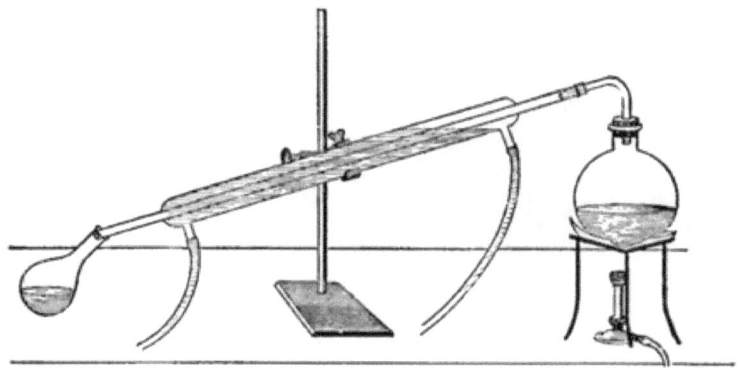

Fig. 21.—Preparation of Chloroform from alcohol and bleaching powder.

EXPT. 7. Mix 50 grams of bleaching powder with 250 c.c. of water, and put the mixture in a large retort, or large flask fitted to a Liebig's condenser (see figure); add 250 c.c. of methylated spirit, and heat the mixture until it begins to boil. Then *remove the burner*, and allow the reaction to proceed by itself. Chloroform and water are condensed in the flask B, the chloroform sinking to the bottom of the water.

The mechanism of this reaction may be explained now, though it will scarcely be fully understood until further acquaintance with the subject has been made. The bleaching powder acts both as an oxidising and as a chlorinating agent. In the first capacity it removes two hydrogen atoms from the ethyl alcohol:

$$CH_3.CH_2.OH + Ca\begin{cases}Cl\\OCl\end{cases} = CH_3.CHO + H_2O + CaCl_2,$$

Ethyl alcohol.   Ethyl aldehyde.

but the product $CH_3.CHO$, ethyl aldehyde, is at the same time chlorinated and converted into trichlorethyl aldehyde or chloral:

$$CH_3.CHO + 3Cl_2 = CCl_3.CHO + 3HCl,$$

Aldehyde.   Chloral.

while the chloral, under the influence of the lime of the bleaching powder, gives chloroform and calcium formate:

$$2CCl_3.CHO + Ca(OH)_2 = 2CHCl_3 + (HCO_2)_2Ca.$$

Chloral.   Chloroform.   Calcium formate.

Chloroform is a heavy liquid of pleasant ethereal smell, and is much used in surgery on account of the property which its vapour possesses of producing insensibility when inhaled.

Carbon Tetrachloride, CCl_4, is the last product of the substituting action of chlorine upon methane:

$$CHCl_3 + Cl_2 = CCl_4 + HCl,$$

and is obtained as a pleasant smelling liquid when boiling chloroform is subjected to the prolonged action of a stream of chlorine gas.

Methyl Iodide, CH_3I, is an important reagent often used in the synthesis of organic compounds. It cannot well be prepared by the direct action of iodine upon methane, but is readily obtained by the action of iodine and phosphorus upon methyl alcohol in the way described in detail for making ethyl iodide. It is a volatile liquid which turns brown when exposed to light; it is much used in the organic laboratory.

Iodoform, CHI_3, is used in surgery on account of its marked antiseptic properties. The method of preparation is analogous to that of chloroform, but instead of bleaching powder, we employ iodine together with some alkali, such as sodium hydrate or carbonate.

EXPT. 8. Dissolve 10 grams of soda crystals in 50 c.c. of water, and add 8 c.c. of methylated spirit. Heat to about 70° C., and then add gradually 5 grams of iodine. Iodoform separates out as a yellow precipitate.

Iodoform is a yellow solid with a characteristic smell, and is slightly soluble in hot water, from which it crystallises in lustrous plates.

Ethyl Chloride, C_2H_5Cl, is a volatile liquid boiling at about 12° C., and can now be obtained sealed up in stout glass tubes, in which it is sold for use as a local anæsthetic in minor surgical operations. It acts in this way by virtue of the intense cold produced by its rapid evaporation when the liquid is allowed to spray from a fine opening upon the part where the operation is to be done.

It is prepared by the action of HCl gas upon ethyl alcohol in the presence of zinc chloride:

$$C_2H_5 . OH + HCl = C_2H_5Cl + H_2O,$$
Ethyl alcohol. Ethyl chloride.

and conversely when heated under pressure with water (better with solution of an alkali) ethyl chloride yields ethyl alcohol:

$$C_2H_5Cl + H_2O = C_2H_5 . OH + HCl.$$

Dichlor-ethane, $C_2H_4Cl_2$, is the second in the series of substitution products obtained by the action of chlorine upon ethane:

$$C_2H_6 + Cl_2 = C_2H_5Cl + HCl,$$
$$C_2H_5Cl + Cl_2 = C_2H_4Cl_2 + HCl, \text{ etc.},$$

but we here meet with a further instance of isomerism, and there are two distinct substances possessing the formula $C_2H_4Cl_2$. In one of these, ethene dichloride, $CH_3 . CHCl_2$, both chlorine atoms are connected with the same carbon atom, while in ethylene dichloride, $CH_2Cl . CH_2Cl$, they are attached one to each of the two carbons.

Ethene Dichloride, $CH_3 . CHCl_2$, is obtained by the action of chlorine upon ethyl chloride, as a rather volatile liquid with a smell similar to that of chloroform.

Ethylene Dichloride, $CH_2Cl . CH_2Cl$, is prepared by the direct combination of ethylene and chlorine, and is the oily liquid from whose formation the old name olefiant gas arose:

$$C_2H_4 + Cl_2 = C_2H_4Cl_2.$$
Ethylene   Ethylene dichloride.

Ethyl Bromide, C_2H_5Br, can be obtained by the action of bromine upon ethane:

$$C_2H_6 + Br_2 = C_2H_5Br + HBr,$$
Ethane.   Ethyl bromide.

but more readily by the action of phosphorus tribromide (or phosphorus and bromine together) upon ethyl alcohol. Phosphorus tribromide reacts with water thus:

$$PBr_3 + 3H_2O = P(OH)_3 + 3HBr,$$

that is to say, the three bromine atoms are exchanged for the same number of hydroxyls. Ethyl (or any other) alcohol behaves similarly to water:

$$PBr_3 + 3C_2H_5OH = P(OH)_3 + 3C_2H_5Br,$$

yielding phosphorous acid and ethyl bromide.

Ethyl bromide is a volatile liquid with a pleasant ethereal smell.

Just as with dichlor-ethane, $C_2H_4Cl_2$, there are also two isomeric substances of the formula $C_2H_4Br_2$, and their modes of formation are precisely similar to those of the corresponding chloro-derivatives.

Ethene Dibromide, CH_3CHBr_2, is obtained by the action of bromine upon ethyl bromide:

$$C_2H_5Br + Br_2 = C_2H_4Br_2 + HBr.$$
Ethene bromide.

Ethylene Dibromide, $CH_2Br \cdot CH_2Br$, by passing a stream of ethylene through bromine contained in a series of gas washing cylinders:

$$C_2H_4 + Br_2 = C_2H_4Br_2$$
Ethylene bromide.

Both ethene and ethylene bromides are heavy liquids with a smell similar to that of chloroform, but they differ markedly in many respects.

Ethyl Iodide, C_2H_5I, is an important reagent prepared in a way precisely similar to that employed for ethyl bromide, except that iodine is used instead of bromine.

EXPT. 9. 10 grams of red phosphorus and 60 c.c. of strong alcohol ("absolute" alcohol must be used; rectified spirits of wine is useless) are placed in a retort, and 100 grams of iodine added little by little. The mixture is allowed to stand for several hours, and is then distilled from the water-bath (see figure). If the alcohol employed has been weak, fumes of

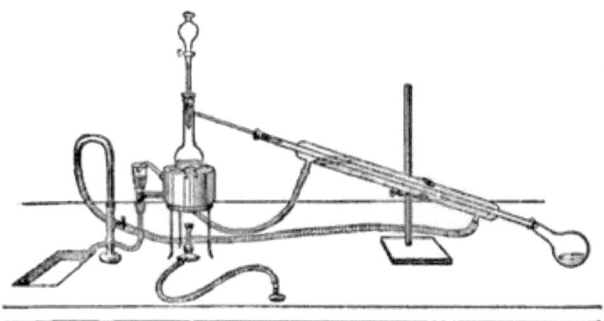

FIG. 22.—Preparation of Ethyl Iodide.

HI will be evolved in torrents, but from absolute alcohol only traces of HI will be given off. The methyl iodide is condensed in the Liebig's condenser, and collects in the flask. It is washed with caustic soda solution and water, then dried by being left to stand in a highly-corked flask over lumps of fused $CaCl_2$, and then re-distilled.

The equation for the reaction is

$$3C_2H_5OH + PI_3 = 3C_2H_5I + H_3PO_3.$$
Ethyl alcohol.   Ethyl iodide.

Ethyl iodide is a colourless liquid with an ethereal smell. It boils at 72°, and is heavier than water, sinking to the bottom like an oil. It gradually decomposes when exposed to light, and the liberated iodine colours the liquid brown. Both methyl and ethyl iodides are largely used in experimental organic chemistry, their use depending on the great mobility of the iodine contained in them. This iodine is readily exchanged for various atoms or radicles by appropriate reactions, and many new compounds have been obtained by this means.

Questions on Chapter V

1. Describe the preparation of chloroform. In what way would you attempt to prove the correctness of the formula $CHCl_3$, which is assigned to it?

2. For what purposes are methyl and ethyl iodides employed, and how are they made?

3. Give an account of the two isomeric substances corresponding to the formula $C_2H_4Br_2$.

CHAPTER VI

THE ALCOHOLS

THE alcohols form a homologous series, of which the starting-point is methyl alcohol, CH_4O. Of the four hydrogens in this molecule one is distinguished from the others by the greater readiness with which it is exchanged for other atoms or radicals, while the fact that methyl alcohol may easily be obtained from or converted into methyl chloride, CH_3Cl, indicates to us that the formula CH_4O may better be written $CH_3.OH$. This leads us to consider water, $H.OH$, as the inorganic type of the alcohols, and it will be useful to remember that there exist many points of resemblance between water and the alcohols in their chemical behaviour.

Methyl alcohol, $CH_3.OH$, being the starting-point of the series, the next member is ethyl alcohol, $C_2H_5.OH$, and so on to the highest known member, myricyl alcohol, $C_{30}H_{61}.OH$; the generic formula is $C_nH_{2n+1}.OH$. Chemically, they are characterised by the presence of the group OH, the hydrogen of which may easily be replaced (1) by sodium or potassium:

$$C_2H_5.OH + Na = C_2H_5.ONa + H,$$
$$\text{Sodium ethylate.}$$

with which compare

$$H.OH + Na = H.ONa + H;$$
$$\text{Sodium hydrate.}$$

or (2) by an acid residue to form an ethereal salt or "ester," in which the alkyl group, C_nH_{2n+1}, takes the place of a monovalent metallic atom in an inorganic salt, as:

(a) $C_2H_5OH + HO\,NO_2 = C_2H_5 \cdot ONO_2 + H_2O$,
    Ethyl nitrate.

(b) $CH_3 \cdot OH + CH_3 \cdot CO_2H = CH_3 \cdot CO_2CH_3 + H_2O$,
    Methyl acetate.

with which compare

(a) $NaOH + HO \cdot NO_2 = NaNO_3 + H_2O$,
    Sodium nitrate.

and (b) $KOH + CH_3 \cdot CO_2H = CH_3 \cdot CO_2K + H_2O$.
    Potassium acetate.

On the other hand, the whole group, OH, in any alcohol is readily driven out by the action either of phosphorus pentachloride or of HCl (in the presence of some hygroscopic substance such as $ZnCl_2$), and its place taken by a chlorine atom:

$CH_3OH + HCl = CH_3Cl + H_2O$
Methyl alcohol.   Methyl chloride.

$C_2H_5OH + PCl_5 = C_2H_5Cl + HCl + POCl_3$.
Ethyl alcohol.   Ethyl chloride.

Methyl Alcohol, $CH_3 \cdot OH$, is contained in wood-spirit, a product of the destructive distillation of wood. In this process, now largely carried on in scientifically-constructed retorts, there are obtained, besides the charcoal left in the retorts, the following: (a) non-condensable gases, chiefly CO, H_2, and CH_4, which, after admixture of hydrocarbon vapour, may be used for illuminating purposes; (b) a watery liquid containing acetic acid, methyl alcohol, and many other substances; and (c) tar. The watery distillate (b) is distilled anew after addition of enough lime to retain the acetic acid, and the crude wood-spirit thus obtained, after some further treatment, is saturated with $CaCl_2$ and heated by steam to 100° C. The impurities are thus driven off, and a residue is left, consisting of a compound of $CaCl_2$ with methyl alcohol. This is mixed with water, when the methyl alcohol is liberated and can be recovered by distillation, but the distillate requires to be again rectified over quicklime in order to free it from water.

Methyl alcohol is a light colourless mobile liquid with a

spirituous odour. When ignited it burns with a pale blue flame. The pure alcohol is used in the manufacture of certain aniline dyes, and for this purpose it should be as free as possible from acetone, a substance largely present in crude wood-spirit; but for other purposes, such as for dissolving resins in the manufacture of varnish, a wood-spirit rich in acetone is desirable, on account of its greater solvent power.

The presence of acetone can be detected and its amount estimated by means of its property of yielding iodoform when treated with iodine and potash. Pure methyl alcohol itself does not produce iodoform, whereas one molecule is obtained from each molecule of acetone present.

Methyl alcohol boils at 66°. It is considerably lighter than water, but mixes with it readily.

Chemically, methyl alcohol is the type of a *primary alcohol*, that is, of one containing the group $-CH_2.OH$. Every primary alcohol when oxidised loses first two atoms of hydrogen, and gives an aldehyde characterised by the group $-CHO$, which can be further oxidised to the group $-COOH$, so yielding an acid. In this particular case the alcohol, $HCH_2.OH$, is first oxidised to formaldehyde, $HCHO$, and this to formic acid, $HCOOH$, by treatment with appropriate oxidising agents.

As in all other alcohols, whether primary or other, the hydrogen of the OH group can be readily replaced by sodium or potassium:

$$2CH_3.OH + 2Na = 2CH_3ONa + H_2,$$
Methyl alcohol.   Sodium methylate.

and the hydroxyl group, as a whole, is substituted by chlorine by the action of PCl_5:

$$CH_3.OH + PCl_5 = CH_3Cl + POCl_3 + HCl,$$
Methyl alcohol.   Methyl chloride.

while, on the other hand, the synthesis of methyl alcohol can be effected by heating methyl chloride with water in sealed tubes to a temperature of 120° C.:

$$CH_3Cl + H.OH = CH_3OH + HCl.$$

Very important also is the ability of methyl alcohol to form ethereal salts, in which the methyl group of the alcohol plays the same part as the metal in an inorganic salt. These are entirely similar to those derived from ethyl alcohol, which will presently be considered in more detail.

Ethyl Alcohol, $C_2H_5.OH$, is prepared on a very large scale, though not in the pure state, by the fermentation of starch or sugar contained in various cereals and fruits. The term fermentation is applied to a process of chemical decomposition, depending for its continuance upon the influence of some "ferment," which yet seems to take no part in the chemical reaction, and is able to transform a disproportionately large amount of the fermenting substance. Ferments may be organised or unorganised. In the first case they are living micro-organisms whose activity as ferments is connected with their vital processes, and ceases with their death. The unorganised ferments are definite chemical substances called enzymes, but no satisfactory explanation has been given of their action.

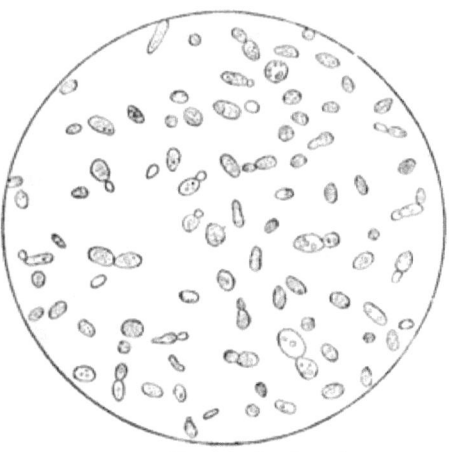

FIG. 23.—Pure Yeast under the microscope.

In the case of the alcoholic fermentation of sugar we are concerned with an organised ferment, the yeast plant. This is a minute and structurally very simple plant, which, placed in a solution of sugar, is able to grow and multiply, provided the temperature be maintained between the limits of 5° and 40° C.; at the same time the sugar is gradually decomposed mainly according to the equation:

$$C_6H_{12}O_6 = 2CO_2 + 2C_2H_6O,$$
$$\text{Glucose.} \qquad\qquad \text{Alcohol.}$$

but there is always produced a certain amount of other substances, of which the most important are higher alcohols and their ethereal salts; these together constitute the *fusel oil* of the crude alcohol, and to its different nature are due both the pleasant flavour of good wine and the foul taste of cheap spirit.

EXPT. 10. Dissolve 10 grams of sugar in 200 c.c. of warm water. Place in a 250 c.c. flask and add some yeast. Fit the flask with a cork and tube to pass any gas evolved through lime-water.

Arrange a parallel experiment, using glucose or honey instead of cane-sugar.

Notice that fermentation soon begins in the latter case, and that CO_2 is evolved. The cane-sugar is much slower.

The most important alcoholic beverages may be classified as follows :—

(a) Beer or ale, made by fermentation of the sugar in malted cereals (especially barley), and containing from 4 to 10 per cent of alcohol.

The malting is itself a fermentation process. The active principle is the *diastase* of the malt, an unorganised ferment which converts the starch of the cereal into sugar.

(b) Wines, made by fermenting the saccharine juice of ripe fruit. No ferment is artificially introduced, as in the case of brewing beer, but micro-organisms floating as dust in the air fall into the must and start fermentation. Wines contain from 10 to 20 per cent of alcohol.

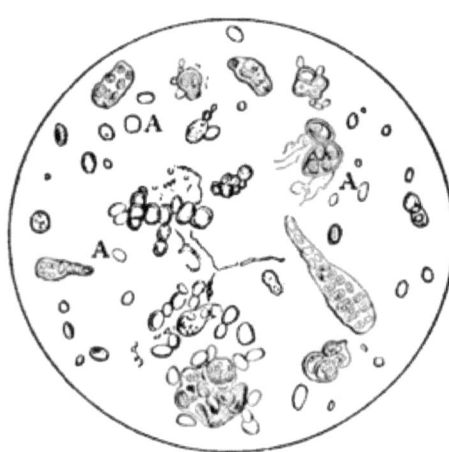

FIG. 24.—Bloom of Grapes under the microscope, showing yeast cells at A.

(c) Spirits are much richer in alcohol (30 per cent and upwards), and are made by distillation of the weaker spirituous beverages. A large quantity of cheap spirit is made by fermentation of malted potato-starch, and from the same source the bulk of the alcohol used in the arts and manufactures is obtained.

Methylated Spirit.—In order that the high price of alcohol due to the heavy duty upon it may not seriously interfere with its use for other purposes than as the basis of intoxicating beverages, the so-called methylated spirit is allowed to be sold duty free. This is a mixture of alcohol containing 20 per cent of water with substances which make the spirit practically undrinkable, but do not seriously interfere with its use for other purposes, and at the same time are difficult to remove. The present regulations order that certain small proportions of wood-spirit and of light petroleum shall be employed as the methylating mixture.

Alcoholometry.—The method in general use for determining the percentage of alcohol present in a given sample of liquid depends upon the gradual variation in the specific gravity of mixtures of alcohol and water as the proportion of alcohol is altered. The following short table illustrates the way in which the specific gravity changes :—

Parts by weight of alcohol in 100 of mixture.	Specific gravity.	Parts by weight of alcohol in 100 of mixture.	Specific gravity.
0	1.000	60	.896
10	.984	70	.872
20	.972	80	.848
30	.958	90	.823
40	.940	100	.794
50	.918	...	...

It would, of course, be quite incorrect to apply this table to such a liquid as beer or wine, in which other substances than water and alcohol are present, and influence the specific gravity. What is done in such cases is to take 100 c.c. of the liquid, distil over two-thirds of it, and make up the distillate to 100 c.c. by adding water; we then have 100 c.c. of liquid containing all the alcohol which was originally present, but freed from the sugar and other non-volatile materials. By taking the specific gravity of this distillate we find at once from the table what percentage of alcohol there was in the beer or wine taken.

The nomenclature employed in this country for stating the results is very complicated. The standard taken is *proof-spirit*, originally spirit of such strength that when poured over gunpowder and set fire to, it was just able to ignite the powder, while a more watery spirit failed to do so. The present legal definition of proof-spirit is that it should be of the specific gravity $\frac{12}{13}$, which corresponds to a proportion of 49.3 parts by weight of pure alcohol in 100 of the mixture. The strength of spirituous liquors is generally expressed as being so many

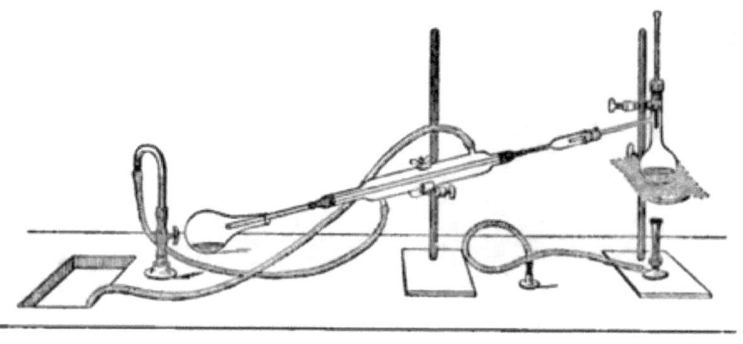

FIG. 25.—Distillation of small quantities.

degrees over or under proof; thirty degrees over proof implies that 100 parts of the spirit contain as much alcohol as 130 parts of proof-spirit.

Pure ethyl alcohol, **absolute alcohol**, is obtained from rectified spirit by treatment with quicklime and subsequent distillation. The spirit is allowed to stand over lumps of quicklime for several hours before distillation, and even then it is usually necessary to repeat the process before the alcohol is entirely freed from water. Whether this is the case can be made certain by shaking some anhydrous copper sulphate (a white powder obtained by heating the crystallised blue salt for several hours at 180° C.) with the alcohol, when the presence of even a trace of water will be detected by the white copper sulphate becoming tinged with blue. Anhydrous alcohol is very hygroscopic, and must be preserved in well-stoppered bottles.

Pure ethyl alcohol has a slight pleasant smell, and boils at a considerably lower temperature than water, viz. 78·3° C. Its specific gravity is .794 at 15° C. It mixes readily with water, and can be used as a solvent for many substances (resins and other organic compounds) which are insoluble in water.

Other methods besides that of fermentation of sugar may be used for the preparation of ethyl alcohol, but are of theoretical interest only ; the most important of them are :—

I. Ethylene, C_2H_4, is absorbed by concentrated sulphuric acid with formation of ethyl-sulphuric acid :

$$C_2H_4 + H_2SO_4 = C_2H_5 . HSO_4,$$

and this, when treated with hot water, is decomposed into alcohol and sulphuric acid :

$$C_2H_5 . HSO_4 + H_2O = C_2H_5OH + H_2SO_4.$$

As ethylene can be prepared by passing a mixture of acetylene and hydrogen over platinum sponge, and acetylene has been made by direct union of carbon and hydrogen, this gives a way by which it would be possible to effect the synthesis of ethyl alcohol from its elements.

II. Ethyl iodide, bromide, or chloride, when heated with water (or more readily, when heated with solution of an alkali) to a high temperature, yields ethyl alcohol,

$$C_2H_5I + H_2O = C_2H_5OH + HI.$$

This last method of preparation is strong evidence for the constitutional formula, $CH_3 . CH_2OH$, which has been adopted ; other evidence is forthcoming in the reactions of ethyl alcohol, all of which are well represented by this formula. Of these reactions the following only will be mentioned here :—

I. With metallic sodium or potassium, an alcoholate is formed and hydrogen is evolved :

$$2C_2H_5 . OH + 2Na = 2C_2H_5ONa + H_2.$$
Sodium ethylate.

The alcoholates are readily oxidised, and are decomposed by water with formation of alcohol and a hydrate :

$$C_2H_5ONa + H_2O = C_2H_5OH + NaOH.$$

II. With PCl_3, or HCl in presence of a dehydrating agent, ethyl chloride is formed:

$$C_2H_5OH + HCl = C_2H_5Cl + H_2O.$$

Zinc chloride may be used as the dehydrating agent. Similar reactions occur with HBr. HI, also PBr_3 and PI_3.

III. With acids the alcohol combines to form ethereal salts (see p. 55):

$$C_2H_5OH + H_2SO_4 = C_2H_5 . HSO_4 + H_2O.$$

Alcohol and sulphuric acid yield ethyl hydrogen sulphate and water.

Propyl Alcohol, C_3H_8O, is the next higher homologue of ethyl alcohol, and is the lowest member of the series for which isomeric forms are possible. If we proceed from ethyl alcohol, $CH_3 . CH_2OH$, by substituting a methyl group, CH_3, for a hydrogen atom, we find that two isomeric alcohols are indicated for the formula C_3H_8O:

$\overset{*}{C}H_3 . CH_2 . OH$ gives

$CH_3 . CH_2 . CH_2 . OH$

Normal propyl alcohol.

$CH_3 . \overset{*}{C}H_2 . OH$ gives

$\genfrac{}{}{0pt}{}{CH_3}{CH_3} {>} \overset{*}{C}H . OH$

Isopropyl alcohol.

and both of these isomers are well-known substances.

A third isomer, $CH_3 . CH_2 . O . CH_3$, exists, but is not an alcohol; it is methyl-ethyl ether.

Normal propyl alcohol is found in fusel oil in considerable quantity, and can also be prepared synthetically by the action of water or potash solution upon the corresponding iodide:

$$CH_3 . CH_2 . CH_2I + H_2O = CH_3 . CH_2 . CH_2OH + HI.$$

Propyl iodide.   Propyl alcohol.

Isopropyl alcohol does not occur in fusel oil, but can only

be obtained by synthetical methods, as by the action of water upon isopropyl iodide:

$$CH_3 . CHI . CH_3 + H_2O = (CH_3)_2CHOH + HI.$$

Both these alcohols are liquids of pleasant smell boiling at a somewhat lower temperature than water. Chemically, both exhibit the reactions characteristic of alcohols which are mentioned on p. 51, but they differ markedly from one another in their behaviour towards oxidising agents. Normal propyl alcohol, when oxidised, yields first an aldehyde and then an acid (propionic):

$$CH_3 . CH_2 . CH_2OH \longrightarrow CH_3 . CH_2 . CHO \longrightarrow CH_3 . CH_2 . CO_2H$$
Primary alcohol.　　　　　Aldehyde.　　　　　Acid.

a behaviour completely parallel to that of methyl and ethyl alcohols, and characteristic of all alcohols containing the group $CH_2 . OH$. Such alcohols are termed **primary alcohols**. Isopropyl alcohol, on the other hand, yields first a ketone, and this, on further oxidation, breaks up into several acids containing a smaller number of carbon atoms in the molecule. This behaviour is characteristic of **secondary alcohols**, which contain the group CHOH united to two alkyl groups:

$$(CH_3)_2 . CHOH \longrightarrow (CH_3)_2 CO \longrightarrow \begin{array}{c} CH_3CO_2H \\ \text{and} \\ H . CO_2H \end{array}$$
Secondary alcohol.　　　Ketone.　　　Lower acids.

The next alcohol we shall consider, butyl alcohol, will furnish us with a case of a **tertiary alcohol**. Such an alcohol contains the group C . OH combined with three alkyl groups, and on oxidation breaks up at once into bodies containing fewer carbon atoms in the molecule:

$$(CH_3)_3 . COH \longrightarrow \text{bodies with fewer carbon atoms in the molecule.}$$
Tertiary alcohol.

Butyl Alcohols, $C_4H_{10}O$, occur in four isomeric forms. Their derivation from the two propyl alcohols by substitution of a methyl group for a hydrogen atom is shown in the following table:

$CH_3 . CH_2 . CH_2OH$ yields (i.) $CH_3 . CH_2 . CH_2 . CH_2OH$
Normal butyl alcohol.

(ii.) $(CH_3)_2CH . CH_2OH$
Isobutyl alcohol.

(iii.) $CH_3 . CH_2 . CH(CH_3) . OH$
Secondary butyl alcohol.

$(CH_3)_2CHOH$ yields (iv.) $CH_3 . CH_2 . CH(CH_3) . OH$

(v.) $(CH_3)_3COH$
Tertiary butyl alcohol.

but of these (iii.) and (iv.) are identical, so that the full number of isomers indicated by theory is four, and this is also the number actually known. Only one of them is sufficiently important to be further mentioned. **Isobutyl alcohol**, $(CH_3)_2CH . CH_2OH$, can be separated from fusel oil, in which it is present, by fractional distillation. It is a liquid boiling at 107° C., and possessing the characteristic smell of fusel oil.

Amyl Alcohol, $C_5H_{12}O$.—For this alcohol there are eight isomers indicated by theory, and of these all are now known. Two of these are largely present in fusel oil, and their mixture is the ordinary "amyl alcohol," a liquid of unpleasant smell, which boils at about 130° C. It is obtained from fusel oil by fractional distillation.

Questions on Chapter VI

1. What tests would you apply to a substance given to you, in order to discover whether it is an alcohol or not?

2. How is methyl alcohol obtained commercially? Mention important points in which it resembles, and others in which it differs from, ethyl alcohol.

3. Give an account of the chief chemical changes which occur in brewing beer from barley. How would you determine the percentage of alcohol in a given sample of beer?

4. What are the characteristics of the three classes of alcohols, primary, secondary, and tertiary?

5. Give an account of the two isomeric alcohols possessing the formula C_2H_8O.

CHAPTER VII

ETHEREAL SALTS

ETHERS—MERCAPTAN

Ethereal Salts.—As has been already mentioned, the alcohols are able to combine with acids somewhat in the same way as the inorganic metallic hydrates. The products in the case of the latter are termed salts, while those formed from the alcohols go by the name of "ethereal salts" or "esters":

$$CH_3.OH + HNO_3 = CH_3NO_3 + H_2O,$$
Methyl nitrate, an ethereal salt.
$$K.OH + HNO_3 = KNO_3 + H_2O,$$
Potassium nitrate, a salt.

There are, however, several differences between the two classes of reactions, of which the most important is that the reaction does not occur so readily or completely with the alcohol as with the base; often, especially when the acid is not one of the strongest, it is necessary to employ some dehydrating agent to favour the reaction.

The reason is that the tendency to the reversed change, such as

$$C_2H_5HSO_4 + H_2O = C_2H_5OH + H_2SO_4,$$

becomes greater as the quantity of water present increases. By combining this water with some hygroscopic substance its effect is diminished.

The most important ethereal salts of ethyl alcohol are perhaps the acetate and acid sulphate.

Ethyl Acetate, $CH_3 . CO_2C_2H_5$, can be prepared by heating a mixture of alcohol and acetic acid with strong sulphuric acid:

$$C_2H_5OH \;+\; CH_3 . CO_2H = CH_3CO_2C_2H_5 + H_2O.$$
Ethyl alcohol.   Acetic acid.   Ethyl acetate.

It is a volatile liquid with a strong fragrant smell. Its formation in the way mentioned is employed as a test for acetic acid or an acetate.

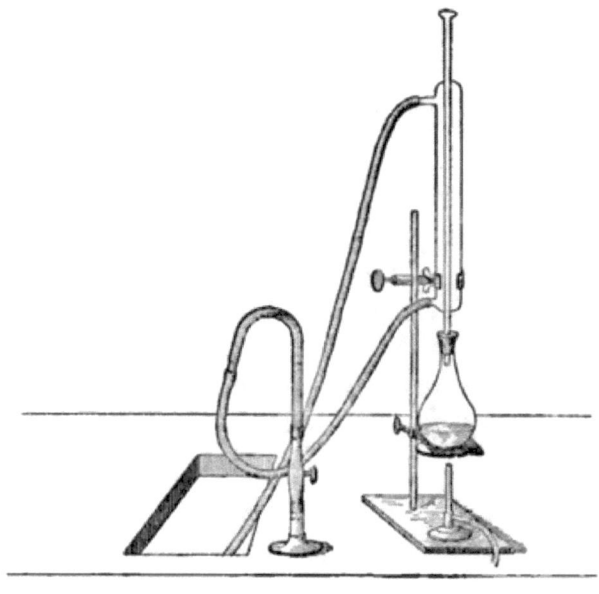

FIG. 26.—Saponification of ethyl acetate by boiling with water and an alkali; the reflux condenser prevents loss by volatilisation.

EXPT. 11. In a test tube mix equal volumes of spirits of wine and strong sulphuric acid. Add some pieces of a solid acetate, and notice the fragrant smell of ethyl acetate which is evolved on gently heating the mixture.

Like other ethereal salts, ethyl acetate is readily split up into the alcohol and acid from which it is formed. The change may be accomplished by heating with water in sealed tubes, or more readily by boiling with a dilute solution of an alkali:

$$CH_3CO_2C_2H_5 + H_2O = CH_3CO_2H + C_2H_5OH.$$
Ethyl acetate.   Acetic acid.   Ethyl alcohol.

Changes of this kind, in which an ethereal salt is broken up by the action of water into the alcohol and acid from which it is formed, are often spoken of as cases of *saponification*. All such changes occur more readily when an alkali or an acid is present in the water.

Ethyl acetate is used in the artificial preparation of perfumes and flavouring essences. Several other ethereal salts, similar in composition, are also used for these purposes, as

Ethyl butyrate, $C_3H_7CO_2C_2H_5$, in pine-apple essence.
Amyl acetate, $CH_3CO_2C_5H_{11}$, in pear essence.

These and similar compounds also constitute the bulk of the natural essences extracted from the plants themselves.

Ethyl Hydrogen Sulphate, $C_2H_5HSO_4$, is formed when alcohol and strong sulphuric acid are mixed. It can be separated from unaltered sulphuric acid by means of its barium salt, which is soluble in water, whereas barium sulphate is insoluble. Ethyl hydrogen sulphate behaves as a monobasic acid, and when liberated from its barium salt, can only be obtained as a thick uncrystallisable syrup.

Two of its reactions are important :—

(*a*) When heated alone it splits up into ethylene and sulphuric acid :

$$C_2H_5HSO_4 = C_2H_4 + H_2SO_4.$$

In this case a larger proportion of sulphuric acid is used, and the temperature of the reaction is higher.

(*b*) When heated with alcohol it forms ether and sulphuric acid :

$$C_2H_5HSO_4 + C_2H_5OH = (C_2H_5)_2O + H_2SO_4.$$

In this case alcohol is present in larger quantity, and the decomposition proceeds at a lower temperature.

ETHER

Ether is now a term applied to a whole class of compounds, all of them oxides of organic radicles, such as methyl and ethyl. Methyl ether is $(CH_3)_2O$, and ethyl ether $(C_2H_5)_2O$. This latter is the ordinary "sulphuric ether" of the chemists, the name being given from the fact that sulphuric acid is used in its manufacture, although the substance obtained has no sulphur whatever in its composition.

Ethyl Ether, $(C_2H_5)_2O$, is ordinarily prepared by the action of ethyl-sulphuric acid upon alcohol:

$$C_2H_5HSO_4 + C_2H_5OH = (C_2H_5)_2O + H_2SO_4.$$

EXPT. 12. In practice a mixture of alcohol and sulphuric acid (consisting, therefore, largely of ethyl-sulphuric acid) is heated in a flask to about 140° C., and then a slow stream of alcohol is allowed to flow into the heated liquid. The vapours given off are condensed, and yield a mixture of water, alcohol, and ether. The layer of ether is separated, dried over quicklime, and redistilled.

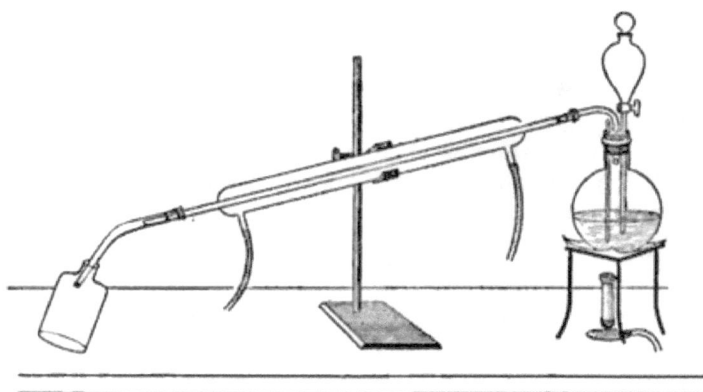

FIG. 27.—Preparation of Ether.

Ethyl ether is a colourless mobile liquid, boiling at 35° C. Its vapour is very heavy, and also readily inflammable, so that care must be exercised in working with ether in the neighbourhood of a flame. The smell of ether is pleasant, but when inhaled in quantity the vapour produces insensibility, and is used as an anæsthetic in cases where chloroform is not permissible on account of its depressing action upon the heart. When drunk in the liquid state ether produces a peculiar kind of short-lived intoxication, the after effects of which are very injurious to the health.

The constitution of ether is not evident from the mode of formation given above, and its reactions are mostly not of a character to throw light upon this point. The preparation from ethyl iodide and sodium ethylate (see p. 51),

$$C_2H_5I + C_2H_5ONa = C_2H_5 . O . C_2H_5 + NaI,$$

is, however, strong evidence in support of the view that ordinary ether is oxide of ethyl.

MERCAPTAN AND ETHYL SULPHIDE

A large number of organic bodies are known in which it seems that an atom of sulphur plays the part of an atom of oxygen in closely related compounds. Such are the *mercaptans* (*e.g.* $C_2H_5 . SH$), which correspond to the alcohols ($C_2H_5 . OH$), and the *alkyl sulphides*, *e.g.* $(C_2H_5)_2S$, which correspond to the ethers ($(C_2H_5)_2O$).

Ethyl Mercaptan, $C_2H_5 . SH$, is formed by the action of potassium sulphydrate, KSH, upon ethyl bromide or iodide:

$$C_2H_5Br + KSH = C_2H_5 . SH + KBr$$
$$c.f. \quad C_2H_5Br + KOH = C_2H_5 . OH + KBr.$$

It is a volatile liquid of very strong and unpleasant odour. The hydrogen of the SH group is more readily replaced by metals than the corresponding H atom in alcohols. Not only does mercaptan react with sodium and potassium, but also with the oxides of heavy metals, such as mercury:

$$HgO + 2C_2H_5 . SH = (C_2H_5S)_2Hg + H_2O,$$

(hence the origin of the name mercaptan, *mercurium aptans*).

Ethyl Sulphide, $(C_2H_5)_2S$, can be prepared,

(1) By acting on potassium sulphide, K_2S, with ethyl bromide :

$$K_2S + 2C_2H_5Br = (C_2H_5)_2S + 2KBr.$$

When a solution of caustic potash is saturated with H_2S, the compound KHS is produced. If to this the same amount as was originally taken of caustic potash solution be added, K_2S is formed :

$$KOH + H_2S = KSH + H_2O$$

and

$$KSH + KOH = K_2S + H_2O.$$

(2) By treating the compound, C_2H_5SK (obtained from mercaptan by action of potassium), with ethyl bromide or iodide:

$$C_2H_5 . SK + C_2H_5Br = (C_2H_5)_2S + KBr,$$

with which compare the method for preparing ether :

$$C_2H_5 . OK + C_2H_5Br = (C_2H_5)_2O + KBr.$$

Ethyl sulphide, like nearly all volatile organic compounds which contain sulphur, has a most unpleasant smell.

QUESTIONS ON CHAPTER VII

1. What is meant by an "ethereal salt"? How are such compounds prepared?

2. Mention some organic compounds of the class of ethereal salts which are used in artificial flavouring essences.

3. What two substances can be prepared by heating ethyl alcohol with sulphuric acid? How do the circumstances of the reaction need to be modified in the two cases.

4. Why is ether regarded as ethyl oxide? What sulphur-containing compound resembles it in composition, and how is it prepared?

CHAPTER VIII

ALDEHYDES AND KETONES

The Aldehydes are characterised by the presence of the monovalent group, CHO, whose structure is represented by the formula $C{\lneq}_H^O$; that is to say, the general behaviour of the aldehydes is best represented by formulæ in which this group is connected with an alkyl group, such as methyl or ethyl.

The aldehydes occupy an intermediate position between the acids and the alcohols by whose oxidation they are produced; thus, between ethyl alcohol, $CH_3 . CH_2OH$, and acetic acid, CH_3COOH, stands the aldehyde $CH_3 . CHO$, or in general:

$R . CH_2OH - H_2 = RCHO$; and $R . CHO + O = R . COOH$;
Alcohol. $\longrightarrow$ Aldehyde. $\longrightarrow$ Acid.

and the names of the aldehydes are best chosen so as to denote their connection with a particular acid, the one into which they are converted by addition of an atom of oxygen. Thus the first member of the aldehyde series, $H . CHO$, is termed formaldehyde, the second one, $CH_3 . CHO$, is acetaldehyde, and so on.

Formaldehyde, $H . CHO$, is best obtained from the corresponding alcohol, methyl alcohol, $H . CH_2OH$, by oxidation; and this is most conveniently effected by passing warm air saturated with the vapour of methyl alcohol over a glowing copper spiral.

EXPT. 13. In the centre of a piece of combustion tubing, about a foot in length, place a two-inch coil of copper gauze. Connect one end of the tube through two gas-washing bottles (the first empty, the second half full of water) with an aspirator, and the other end with a gas-washing bottle containing methyl alcohol, kept at about 50° by being placed in a beaker of warm water.

Now turn on the water tap of the aspirator until a vigorous current of vapour-laden air is passing over the copper gauze. Heat this gently with a Bunsen burner until it begins to glow, when it will continue to do so without any further use of the burner so long as the experiment is continued. In order to minimise the danger of cracking the glass tube when the copper spiral suddenly begins to glow, it is well to support the spiral on a thin piece of mica or of asbestos paper.

In this way are obtained only mixtures of formaldehyde with methyl alcohol and water. It has not been found possible to prepare pure formaldehyde, H . CHO, except in solution or in the state of vapour. When the solution is evaporated or the vapour cooled, a solid substance is obtained of the same composition as formaldehyde, but not of the same molecular weight; this is *para-formaldehyde*, and has the formula $(CH_2O)_x$, where x is possibly equal to 3, but is not known with certainty. Para-formaldehyde, $(CH_2O)_3$ (?), is said to be *polymeric* with formaldehyde, CH_2O, as the two substances have the same composition, but different molecular weights. The opposite change is easily accomplished by vapourising the solid para-formaldehyde when a vapour whose density shows it to be made up of the simple molecules, CH_2O, is obtained; but on cooling, these again gradually unite to the more complex molecules, $(CH_2O)_3$.

Beyond its tendency to polymerisation, the chief characteristic of formaldehyde is the readiness with which it takes up oxygen from other substances to effect the change into formic acid :

$$H . CHO + O = H . COOH ;$$
Formaldehyde.   Formic acid.

accordingly, formaldehyde is a strong reducing agent; it reduces in the cold both Fehling's solution and solutions of silver salts.

EXPT. 14. Prepare a quantity of Fehling's solution by dissolving 100 grams of Rochelle salt (sodium potassium tartrate) in a little water, adding 30 grams of NaOH in 300 c.c. of water, and then 20 grams of crystallised $CuSO_4$, dissolved in about 100 c.c. of water; mix and keep in a stoppered bottle.

Place some of the formaldehyde solution prepared in Expt. 13 in a beaker, and add Fehling's solution drop by drop. Notice that its dark blue colour is discharged and a light red precipitate produced. This is Cu_2O, formed by reduction of the $CuSO_4$:

$$2CuSO_4 + 4KHO + H.CHO = HCO_2H + Cu_2O + 2K_2SO_4 + 2H_2O.$$

EXPT. 15. Dissolve 3 grams $AgNO_3$ in a mixture of 20 c.c. strongest ammonia solution (sp. gr. .88) with its own volume of water, and add a solution of 3 grams NaOH in 25 c.c. of water. Keep in a small stoppered bottle in a dark place.

Take some of this silver solution in a test tube, and add a few drops of the formaldehyde solution; allow to stand in the cold. In a few minutes a brilliant mirror of metallic silver will be deposited on the sides of the test tube.

$$2AgNO_3 + H_2O + HCHO = HCOOH + 2HNO_3 + 2Ag.$$

Acetaldehyde, $CH_3.CHO$, is prepared by the oxidation of ethyl alcohol by distillation with a mixture of potassium bichromate and dilute sulphuric acid.

EXPT. 16. Place in a flask 30 grams $K_2Cr_2O_7$ (in small lumps) and 120 c.c. water. Mix in a beaker 40 c.c. methylated spirit and 25 c.c. strong sulphuric acid, and allow to cool. Then add this mixture gradually to the bichromate, taking care to keep cool by running water over the outside of the flask.

Heat the mixture on the water-bath, and collect the distillate in a receiver kept cold by ice. The impure acetaldehyde collected can be purified partly by fractional distillation, and finally by conversion into the solid compound which it forms with ammonia.

Acetaldehyde is a colourless liquid, boiling at 21° C. and possessing a characteristic smell. Like formaldehyde, it is a strong reducing agent, as may be shown by experiments similar to Nos. 14 and 15. It further resembles the lower member of the aldehyde series in the readiness with which it polymerises to *paraldehyde* $(C_2H_4O)_3$. Acetaldehyde itself is a colourless very volatile liquid (B.P. 21° C.) with a pleasant smell, but on standing in contact with even a trace of various substances—H_2SO_4, HCl, SO_2, etc.—it changes almost entirely to paraldehyde, which is a liquid at ordinary temperatures, but solidifies in a freezing mixture, boils at 124°, and gives a vapour whose density corresponds to the molecular formula $(C_2H_4O)_3$. Paraldehyde, though so directly obtained from acetaldehyde, is not itself a real aldehyde at all. This is shown by its whole chemical behaviour, especially by the fact that it does not reduce metallic silver from an ammoniacal silver solution, and leads us to conclude that in the formula of paraldehyde the group $-CHO$ no longer occurs. Paraldehyde is easily reconverted into ordinary acetaldehyde by distillation with a little H_2SO_4.

Metaldehyde is another substance of the formula $(C_2H_4O)_3$ obtained by polymerisation of acetaldehyde; it is isomeric with paraldehyde.

Compounds of Acetaldehyde.—In some respects acetaldehyde is more typical than its lower homologue of the group of aldehydes, and we have therefore delayed till now the consideration of certain reactions exhibited by aldehydes as a class, in which certain substances, such as NH_3, HCN, etc., are added to the aldehyde molecule.

(a) Acetaldehyde, like all the other aldehydes except $H.CHO$, unites directly with ammonia to form a compound of the type $R.CH(OH)(NH_2)$:

$$CH_3.CHO + NH_3 = CH_3.CH{<}^{OH}_{NH_2}.$$

This particular one is called simply aldehyde-ammonia, and is formed as a white crystalline solid when dry NH_3 is passed into an ethereal solution of aldehyde. It is decomposed by dilute acids into aldehyde and ammonia.

EXPT. 17. Pass NH_3, dried by quicklime, into a solution of aldehyde in ether; collect on a filter the white precipitate produced, and show that some of it when warmed with dilute H_2SO_4 regenerates aldehyde.

(b) Addition compounds with HCN are also formed by the aldehydes, thus

$$CH_3.CHO + HCN = CH_3.CH{<}^{OH}_{CN}.$$

(c) Sodium hydrogen sulphite, $NaHSO_3$, also gives addition products, which are often used as a means of separating and purifying the aldehydes. The following equation represents what happens in the case of acetaldehyde:

$$CH_3.CHO + NaHSO_3 = CH_3.CH{<}^{OH}_{SO_3Na}.$$

Compounds of this type are obtained when an aldehyde is shaken with a saturated solution of $NaHSO_3$. They are white crystalline solids, soluble in water, and decomposed by dilute acids with regeneration of the aldehyde.

Chloral is a very important derivative of ordinary aldehyde; its formula is $CCl_3.CHO$, and its systematic name *trichlor-aldehyde*.

Chloral is prepared by passing chlorine into alcohol and decomposing the solid crystalline product (a compound of chloral and alcohol) with sulphuric acid. The reaction may be represented as occurring in two parts:

(*a*) The alcohol is oxidised to aldehyde,

$$CH_3.CH_2OH + Cl_2 = CH_3.CHO + 2HCl.$$
Ethyl alcohol.   Acetaldehyde.

(*b*) The aldehyde is converted into trichlor-aldehyde or chloral,

$$CH_3.CHO + 3Cl_2 = CCl_3.CHO + 3HCl.$$
Acetaldehyde.   Chloral.

It is a liquid with a penetrating smell, and possesses most of the properties (reducing power, etc.) characteristic of the aldehydes. It is decomposed by alkalies with production of chloroform:

$$CCl_3.CHO + KOH = CHCl_3 + H.CO_2K,$$
Chloral.   Chloroform. Potassium formate.

hence perhaps the well-known narcotic power of chloral.

Chloral Hydrate, $CCl_3.CHO + H_2O$, is a compound of chloral with water, produced by direct combination of the two liquids. It is a crystalline solid, and is the form in which chloral is usually administered.

KETONES

The ketones are a series of compounds resembling in many respects the aldehydes, but differing in others; and we attempt to represent both resemblances and differences by giving to the ketones the formula $R.CO.R$, closely allied to the aldehyde formula $R.CO.H$.

F

(1) Just as the aldehydes are obtained by carefully graduated oxidation of primary alcohols,

$$R \cdot CH_2OH + O = R \cdot CO \cdot H + H_2O,$$
Primary alcohol. Aldehyde.

so the ketones are the first products formed by the oxidation of secondary alcohols, that is, alcohols in which the group CH(OH) is combined with two alkyl groups:

$$\genfrac{}{}{0pt}{}{R}{R'}{>}CH \cdot OH + O = \genfrac{}{}{0pt}{}{R}{R'}{>}CO + H_2O.$$
Secondary alcohol. Ketone.

(2) Another method of general application for the preparation of ketones is the dry distillation of the calcium salts of fatty acids; thus calcium acetate gives acetone or di-methyl ketone:

$$\genfrac{}{}{0pt}{}{CH_3CO \cdot O}{CH_3COO}{>}Ca = \genfrac{}{}{0pt}{}{CH_3}{CH_3}{>}CO + CaCO_3.$$
Calcium acetate. Acetone.

(3) A third method of considerable importance is the treatment of acetyl chloride or similar compounds with zinc methyl, ethyl, etc. The reaction may be represented thus:

$$2CH_3COCl + Zn(C_2H_5)_2 = 2CH_3COC_2H_5 + ZnCl_2.$$
Acetyl chloride. Zinc ethyl. Methyl-ethyl ketone.

The ketones resemble the aldehydes in their power of forming addition products with HCN, and with $NaHSO_3$. They do not possess the same energetic reducing power, nor do they combine with ammonia in the same way as the aldehydes.

Acetone, $(CH_3)_2CO$, or dimethyl ketone is the simplest ketone. It can be prepared by any of the general methods given above, the one generally adopted being the dry distillation of calcium acetate:

$$(CH_3CO_2)_2Ca = (CH_3)_2CO + CaCO_3.$$

Acetone is also found amongst the products of the dry distillation of wood (see p. 45), and is largely obtained from that source. It is used as a solvent, and for the preparation of iodoform and other substances. It is a volatile inflammable liquid with a pleasant smell.

Oximes and Hydrazones.—Special importance attaches to the compounds which aldehydes and ketones form with hydroxylamine, $NH_2(OH)$, and phenylhydrazine, $C_6H_5NH.NH_2$. In these oximes and hydrazones the oxygen of the CO group in the aldehyde or ketone is replaced by a divalent residue, thus:

$$(CH_3)_2CO + H_2N.OH = (CH_3)_2C:N.OH + H_2O$$
$$\text{Oxime.}$$

$$CH_3COH + H_2N.NHC_6H_5 = (CH_3)CH:N.NHC_6H_5 + H_2O.$$
$$\text{Hydrazone.}$$

The importance of these oximes and hydrazones lies in their great utility as a means of characterising the various aldehydes and ketones. The hydrazones especially are usually crystalline solids, only slightly soluble in the ordinary solvents, and are therefore much more easily identified than the aldehydes or ketones from which they are prepared.

QUESTIONS ON CHAPTER VIII

1. How is acetaldehyde prepared? Mention its chief properties.
2. How are the aldehydes as a class characterised by their reactions, and how is their behaviour represented in the generic formula $R.CHO$?
3. What is the relation of chloral to acetaldehyde? Give its preparation and properties.
4. What bodies are formed by the oxidation of (*a*) ethyl alcohol, (*b*) aldehyde, (*c*) chloral?
5. Illustrate the chief points of resemblance and difference in the chemical behaviour of aldehyde and acetone.

CHAPTER IX

THE FATTY ACIDS

The Fatty Acids form an important homologous series, some higher members of which are contained in all natural fats. The lower members are liquids of strongly acid character and sharp penetrating odour, but with increasing molecular weight the members of the series lose their solubility in water, and with it their acid taste and power of turning blue litmus red. The power of forming salts is, however, unimpaired even in the highest member of the series yet obtained.

The first acid of the series is formic acid, CH_2O_2, the second acetic acid, $C_2H_4O_2$. The general formula of the whole series is $C_nH_{2n}O_2$, but this is better written $C_nH_{2n+1}.CO_2H$, to indicate that every acid of the series contains the "carboxyl" group, CO_2H, combined with a hydrocarbon residue (or "alkyl" group), such as methyl, CH_3, ethyl, C_2H_5, etc. Formic acid is then written $H.CO_2H$, acetic acid $CH_3.CO_2H$, and so on.

The reasons for writing the formulæ in this way will be best understood if we consider in detail the case of acetic acid; this has the molecular formula $C_2H_4O_2$. Of the four hydrogens only one can be replaced by metals, *i.e.* the acid is monobasic, and therefore one of the four hydrogen atoms is differently related to the molecule from the other three. Again the action of phosphorus pentachloride on acetic acid or on sodium acetate yields a substance, acetyl chloride, of the formula C_2H_3OCl; that is, a Cl atom takes the place of an O atom and an H atom. This could not happen unless that O and that H were connected to form the monovalent hydroxyl group OH.

We have now arrived at the formula, $C_2H_3O \cdot OH$, for acetic acid. The next question is whether the three remaining hydrogen atoms are all connected to the same carbon atom or not. Acetic acid treated with chlorine yields a derivative trichloracetic acid of the formula $C_2Cl_3O \cdot OH$, in which the hydroxyl group is still present, and the other three hydrogens are replaced by chlorine. Now trichloracetic acid readily yields chloroform when boiled with water:

$$C_2Cl_3O \cdot OH = CHCl_3 + CO_2,$$
Trichloracetic acid. Chloroform.

showing that all three Cl atoms, and therefore the three hydrogen atoms, whose places they occupy, are connected to the same carbon. Hence acetic acid contains the groups CH_3 and OH, and the only formula in agreement with these experimental results is $CH_3 \cdot COOH$.

General Methods of Preparation.—(1) The first method is one which also furnishes valuable evidence in favour of the formula $C_nH_{2n+1} \cdot CO_2H$ for the series, inasmuch as we start in each case from a substance, $C_nH_{2n+1} \cdot CN$, in order to prepare the corresponding acid. Such an alkyl cyanide is obtained by treating the iodide of the same radical with silver cyanide, *e.g.*:

$$CH_3I + AgCN = CH_3 \cdot CN + AgI,$$
Methyl iodide.   Methyl cyanide.

and when heated with water undergoes a reaction of the following type:

$$CH_3 \cdot CN + 2H_2O = CH_3 \cdot CO_2H + NH_3.$$
Methyl cyanide.   Acetic acid.

Such a reaction is spoken of as hydrolysis, and takes place much more readily when a dilute mineral acid is used instead of pure water; or a solution of an alkali may be employed.

(2) By the action of carbon monoxide on the sodium compound of an alcohol, *e.g.* from sodium methylate sodium acetate is obtained:

$$CH_3 \cdot ONa + CO = CH_3 \cdot COONa.$$

Similarly, sodium formate may be obtained from sodium hydrate:

$$H \cdot ONa + CO = H \cdot COONa.$$

This method is of theoretical interest only.

3. An important practical method is the oxidation of a primary alcohol containing the same number of carbon atoms as the acid to be prepared :

$$CH_3 \cdot CH_2OH + O_2 = CH_3 \cdot COOH + H_2O.$$
$$\text{Ethyl alcohol.} \qquad\qquad \text{Acetic acid.}$$

In the laboratory a mixture of potassium bichromate with dilute sulphuric acid is usually employed as the oxidising agent ; in the commercial manufacture of acetic acid (vinegar) the oxygen of the air is utilised.

In this method the group $CH_2(OH)$ is oxidised to $COOH$. When a less complete oxidation is effected, the product is an aldehyde containing the group CHO :

$$R \cdot CH_2(OH) \longrightarrow R \cdot CHO \longrightarrow R \cdot COOH.$$
$$\text{Primary alcohol.} \qquad \text{Aldehyde.} \qquad \text{Acid.}$$

Formic Acid, $H \cdot CO_2H$, may be prepared by any of the three general methods, *i.e.* :

(1) From HCN, hydrocyanic acid, by heating with a dilute mineral acid in sealed tubes :

$$HCN + 2H_2O = H \cdot CO_2H + NH_3.$$
$$\text{Formic acid.}$$

(2) From sodium or potassium hydrate, and carbon monoxide :

$$NaOH + CO = H \cdot COONa,$$
$$\text{Sodium formate.}$$

a reaction which occurs with great readiness when moist CO is passed over porous soda-lime heated to about 200° C.

(3) By the oxidation of methyl alcohol :

$$HCH_2 \cdot OH + O_2 = HCOOH + H_2O.$$
$$\text{Methyl alcohol.} \qquad \text{Formic acid.}$$

Another method of considerable interest in connection with the physiological chemistry of plants by which formates can be obtained is by the reduction of CO_2 in the presence of water. Thus, when thin slices of metallic potassium are exposed to a moist atmosphere of CO_2 they are gradually converted into potassium formate and carbonate :

$$2K + 2CO_2 + H_2O = H . CO_2K + KHCO_3.$$

Possibly this reduction to formic acid is the first step in the transformation by plants of CO_2 into sugar and starch.

The most practically useful method for preparing formic acid is by the decomposition of oxalic acid. This, when heated alone, or better with glycerine, breaks up as follows :

$$C_2H_2O_4 = H . CO_2H + CO_2.$$
Oxalic acid. Formic acid.

A mixture of equal quantities of glycerine and crystallised oxalic acid is heated in a retort until no more CO_2 is evolved. The distillate collected during this period is a very weak formic acid. On adding more oxalic acid, and again heating, a stronger acid will be obtained, but the acid got in this way never contains less than about 40 per cent of water.

Anhydrous formic acid is prepared from lead formate by the action of hydrogen sulphide. The lead salt is easily obtained from any (weak) formic acid. It is dried and then exposed to a stream of H_2S gas in a tube kept warm by means of a steam jacket. The anhydrous acid distils over, and is a colourless liquid with an acrid odour and very caustic properties.

Formic acid differs from the other members of the series in being a strong reducing agent. It reduces solutions of silver and mercury salts with separation of the metals. When heated with strong sulphuric acid it is decomposed into CO and water :

$$H . CO_2H = H_2O + CO.$$

The **Formates** of the alkali metals are fairly stable substances, which crystallise only with difficulty. Those of the heavy metals, such as silver, are very easily decomposed with separation of the metal.

Acetic Acid, $CH_3 . CO_2H$, can be obtained by any of the three general methods, *i.e.*:

(1) From $CH_3.CN$, methyl cyanide, by heating with a dilute mineral acid or a dilute alkali :

$$CH_3.CN + 2H_2O = CH_3.CO_2H + NH_3.$$
Methyl cyanide. Acetic acid.

Methyl cyanide, CH_3CN, can be made by acting with methyl iodide on silver cyanide :
$$CH_3I + AgCN = CH_3CN + AgI.$$

(2) From sodium or potassium methylate and carbon monoxide :
$$CH_3.ONa + CO = CH_3.COONa.$$
Sodium methylate. Sodium acetate.

For sodium methylate, see p. 46.

(3) By the oxidation of ethyl alcohol :
$$CH_3CH_2OH + O_2 = CH_3.COOH + H_2O.$$
Ethyl alcohol. Acetic acid.

The only one of these methods employed on a large scale is the third. In the preparation of vinegar (which is a dilute acetic acid flavoured by minute quantities of other substances) the alcoholic liquid, whether wine, diluted brandy, or merely potato-spirit and water, is exposed to the simultaneous action of the air (which supplies the oxygen), and of the fermentative influence of a particular organism, the *mycoderma aceti*. The process is carried on most rapidly by allowing the alcoholic liquor to trickle through tubs filled with shavings,

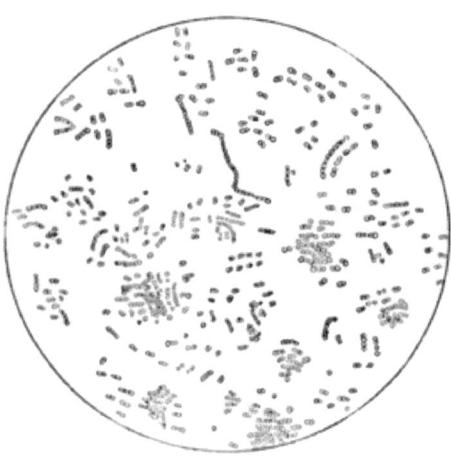

FIG. 28.—The *mycoderma aceti* or "mother of vinegar" seen under the microscope.

on which the mycoderma has developed. New shavings are

at first almost inactive, but they soon become coated with the organism, and the oxidation then takes place readily.

Large quantities of acetic acid are also obtained as one of the products of the destructive distillation of wood. The acid is separated from the other products, chiefly methyl alcohol and acetone, by neutralising with lime and distilling the alcohol and acetone from the calcium acetate. This last is then decomposed by addition of sulphuric acid, and the acetic acid recovered by distillation.

Acetic acid, when perfectly free from water, is a crystalline solid which melts at 17° C. The strongest acetic acid of commerce is termed "glacial acetic acid," from the fact that it is solid in moderately cold weather. It has a penetrating acid smell, and acts like a caustic on the skin.

The salts of acetic acid, the *acetates*, are prepared by acting with the acid upon the oxide or carbonate of the metal whose acetate is required. They are all more or less readily soluble in water, and crystallise well.

Sodium Acetate, $NaC_2H_3O_2$, crystallises with three molecules of water. When heated above 100° C. the water of crystallisation is driven off, and the anhydrous sodium acetate is left as an amorphous mass. The anhydrous salt is used in organic synthesis as a dehydrating agent, and in the preparation of methane:

$$NaC_2H_3O_2 + NaOH = CH_4 + Na_2CO_3.$$

Ammonium Acetate, $(NH_4)C_2H_3O_2$, is a deliquescent solid. Its solution is made use of in qualitative analysis for dissolving lead sulphate, and so separating it from mercuric sulphide, with which it may be mixed in the course of working through the second group of metals.

When strongly heated, ammonium acetate yields acetamide and water:

$$CH_3COONH_4 = CH_3CONH_2 + H_2O.$$

Calcium Acetate, $Ca(C_2H_3O_2)_2$, is used for preparing acetone:

$$\begin{matrix} CH_3COO \\ CH_3COO \end{matrix} \!\!>\!\! Ca = \begin{matrix} CH_3 \\ CH_3 \end{matrix} \!\!>\!\! CO + CaCO_3.$$

Lead Acetate, $Pb(C_2H_3O_2)_2$, is the "sugar of lead" of commerce, and is made by dissolving litharge in acetic acid. It is largely used in the manufacture of white-lead (basic lead carbonate) and chrome-yellow ($PbCrO_4$).

Aluminium Acetate is obtained in solution when calcium acetate is mixed in the presence of water with aluminium sulphate. The solution decomposes on evaporation into acetic acid (which escapes as vapour) and alumina; hence the extensive use of aluminium acetate as a mordant, the alumina combining with the dye to form an insoluble lake, which adheres firmly to the fibre of the cloth.

As tests for the presence of acetic acid may be utilised either the dark red colour of the solution of ferric acetate (destroyed on boiling, with separation of a basic acetate of iron) formed when ferric chloride is added to a solution of an acetate, or the formation of ethyl acetate with its characteristic pleasant smell when an acetate is heated with alcohol and concentrated sulphuric acid.

It is, however, to be noted that these tests give almost identical results with any of the acids of this series. To distinguish acetic acid from its higher homologues, the surest plan is to prepare the silver salt and determine the percentage of silver which it contains.

Propionic Acid, $C_2H_5 \cdot CO_2H$, may be prepared by any of the three general methods, but most conveniently by the third, starting from normal propyl alcohol:

$$C_2H_5 \cdot CH_2OH + O_2 = C_2H_5 \cdot COOH + H_2O.$$
Propyl alcohol.   Propionic acid.

The constitution of propionic acid is shown by this method of preparation, as also by that from ethyl cyanide by hydrolysis:

$$C_2H_5CN + 2H_2O = C_2H_5CO_2H + NH_3.$$
Ethyl cyanide.   Propionic acid.

Butyric Acid, $C_3H_7 \cdot CO_2H$, is the lowest member of the series for which isomeric forms are theoretically possible or have been actually obtained. These are two in number, viz.:

(i) Normal butyric acid, $CH_3CH_2CH_2CO_2H$.

(ii) Isobutyric acid, $\begin{matrix}CH_3\\CH_3\end{matrix}\!>\!CHCO_2H$.

Their constitution is made clear by their synthetical preparation from normal propyl iodide and isopropyl iodide respectively through the intermediary of the cyanides:

(i) $CH_3CH_2CH_2I \longrightarrow CH_3CH_2CH_2CN \longrightarrow CH_3CH_2CH_2CO_2H$
Normal propyl iodide.   Normal propyl cyanide.   Normal butyric acid.

(ii) $(CH_3)_2CHI \longrightarrow (CH_3)_2CHCN \longrightarrow (CH_3)_2CHCO_2H$
  Isopropyl iodide.   Isopropyl cyanide.   Isobutyric acid.

Normal Butyric Acid, $C_3H_7 . CO_2H$, is present in butter in the form of glycerine butyrate, the ethereal salt of glycerine and butyric acid. There are, however, many other similar compounds of glycerine contained in butter, and the isolation of the butyric acid in a state of purity is a matter of difficulty, and the acid is more cheaply and easily obtained by the fermentation (under the influence of the *bacillus subtilis*) of sugar or starch. This process is carried out on a fairly large scale, the butyric acid being converted into its ethyl salt, which is used as a flavouring under the name of essence of pine-apples. The same fermentation of starch and sugar occurs in the human stomach in certain cases of deranged digestion. Starting from glucose (see Chapter XVIII), the change produced by the fermentation may be represented by the equation:

$$C_6H_{12}O_6 = C_4H_8O_2 + 2CO_2 + 2H_2.$$
  Glucose.   Butyric acid.

Butyric acid is an oily liquid with an unpleasant rancid smell. The change which butter undergoes in turning rancid may be represented (so far as the glycerine butyrate in it is concerned) as follows:

$$C_3H_5(OCOC_3H_7)_3 + 3H_2O = C_3H_5(OH)_3 + 3C_3H_7 . CO_2H,$$
  Glycerine butyrate.   Glycerine.   Butyric acid.

and it is to the presence of free butyric acid in rancid butter that its characteristic taste and smell are due.

Isobutyric Acid, $C_3H_7CO_2H$, can be prepared from isopropyl cyanide (see above), and resembles the normal acid in smell and taste, though differing considerably from it in some other physical and chemical properties.

The fifth acid of the series is valerianic acid, $C_4H_9 . CO_2H$, and theory indicates the possible existence of four isomers, all of which have been actually prepared. They may all four be regarded as derivatives of acetic acid, obtained from it by replacing one or more of the hydrogens of its methyl group by alkyl radicles. They are

(i) Propyl-acetic acid, $C_3H_7 . CH_2CO_2H$.
(ii) Isopropyl-acetic acid, $C_3H_7 . CH_2CO_2H$.
(iii) Ethyl-methyl-acetic acid, $\genfrac{}{}{0pt}{}{CH_3}{C_2H_5} > CH . CO_2H$.
(iv) Trimethyl-acetic acid $(CH_3)_3C . CO_2H$.

Of these the second is present in valerian root, while all of them may be prepared synthetically by utilising some one or other of the general methods given on p. 69 as applicable to all the fatty acids.

The higher acids are not of great importance until we come to those with sixteen and eighteen atoms of carbon in the molecule.

Palmitic Acid, $C_{15}H_{31} . CO_2H$, and **Stearic Acid,** $C_{17}H_{35} . CO_2H$, occur in combination with glycerine as the chief constituents of animal fats. Glycerine palmitate is also largely present in most vegetable oils. These fats and oils serve as the starting-point of three important manufactures— those of soap, glycerine, and stearic acid. In making soap the fats are heated with caustic potash or caustic soda solution, and in this way the compounds of glycerine with various fatty acids which are contained in the fat are "saponified," that is, are converted into glycerine and the potassium or sodium salts of the acids. If we consider only one of these compounds, the glycerine stearate, the change may be represented thus :

$$C_3H_5 \begin{cases} OCOC_{17}H_{35} \\ OCOC_{17}H_{35} \\ OCOC_{17}H_{35} \end{cases} + 3NaOH = C_3H_5(OH)_3 + 3C_{17}H_{35}COONa.$$

Glycerine stearate   Glycerine.   Sodium stearate
or fat.   or soap.

The other compounds present undergo similar changes, so that there is obtained along with glycerine a mixture of the palmitates and stearates of sodium or potassium. *This mixture constitutes soap, hard soap being the sodium salts and soft soap the potassium salts of the acids present in the fats employed.* Formerly it was the practice for each household to make its own soap, but the tendency of modern life has led to the manufacture being carried on almost entirely in very large works. The process is of the simplest. The fat or oil is heated to boiling in large open vats along with the solution of soda or potash. When saponification is complete common salt is added, in order to make the soap separate from its solution in the mixture of water and glycerine, and a mass of genuine "curd soap" is obtained in the solid state above the liquid of dilute glycerine. In many soap-boiling establishments the practice is adopted of allowing the mixture of soap, glycerine, and water, which is the immediate result of the boiling with alkali, to solidify together. This kind of soap can obviously be sold at a much lower rate than the genuine product, but is also far less durable.

In working up the fats for glycerine and stearic acid the process is again one of "saponification," that is, splitting up an ethereal salt of glycerine into glycerine and the acid combined with it:

$$\text{Fat} + \text{water} = \text{glycerine} + \text{stearic, etc., acid,}$$

and is now most largely carried on by subjecting the fat to the action of superheated steam in the presence of water and a small proportion of lime. The product is freed from lime by adding the proper amount of sulphuric acid; and in this way the whole of the acids present in the fat are obtained in the form of a semi-solid mass, consisting chiefly of *palmitic acid*, $C_{15}H_{31} . CO_2H$, *stearic acid*, $C_{17}H_{35} . CO_2H$, and an unsaturated acid (see p. 28), *oleic acid*, $C_{17}H_{33} . CO_2H$. Of these the first two are solid, while the last is liquid at ordinary temperatures, and can be removed by applying hydraulic pressure to the mixture of the three acids. The residue is the "stearic acid" of commerce, and is largely employed in the manufacture of candles.

The isolation of pure palmitic or stearic acid from this mixture is a matter of some difficulty, and can only be accomplished by a tedious process of fractional precipitation and crystallisation. The two acids are very similar in their properties, but differ in their melting points.

The process of fractional precipitation is carried out by adding to the solution of the two acids in alcohol a quantity of magnesia, only sufficient to combine with about a third of the amount of acid present. The precipitate is found to contain a much larger proportion of magnesium stearate than the original mixture did of stearic acid, while the palmitic acid is almost entirely left in solution.

The *salts of palmitic and stearic acids* are all devoid of any tendency to crystallise. A mixture of the sodium salts constitutes the essential portion of hard soap, while soft soap contains the potassium salts. The lead salts are also important; they are obtained by treating fats or oils with latharge (lead oxide) and water, and form what is known as lead plaster.

QUESTIONS ON CHAPTER IX

1. The analysis of acetic acid leads to the empirical formula CH_2O. Give reasons for adopting the molecular formula $C_2H_4O_2$.

2. Give arguments supporting the structural formula $CH_3 . CO_2H$ for acetic acid.

3. Write down the formulæ and names of the first four of the series of fatty acids. Give three general methods which (with the necessary modifications) may be applied to the preparation of each of them.

4. Mention the most convenient ways of obtaining formic and acetic acids in quantity. Point out any important difference between the two homologous acids.

5. Give an account of the two isomeric butyric acids, their constitution, preparation, and properties.

6. What is the chemical constitution of fats? How is soap made from them?

CHAPTER X

ACETYL CHLORIDE AND ACETIC ANHYDRIDE

ACETIC acid, CH_3COOH, may be regarded as the hydrate of the radicle CH_3CO to which the name "acetyl" is given. The chloride and oxide of acetyl are of great importance as reagents in the organic laboratory.

Acetyl Chloride, CH_3COCl, is prepared by treating acetic acid with phosphorus trichloride, and distilling the mixture :

$$3CH_3COOH + PCl_3 = 3CH_3COCl + P(OH)_3$$
Acetic acid.   Acetyl chloride.   Phosphorous acid.

It is thus obtained as a colourless volatile liquid, which fumes strongly in the air, and combines eagerly with water to form hydrochloric and acetic acids :

$$CH_3CO|Cl + H|OH = HCl + CH_3COOH.$$
Acetyl chloride.   Acetic acid.

Exactly parallel is the action of acetyl chloride upon organic alcohols and similar bodies which contain the group hydroxyl, OH. Thus with ethyl alcohol it gives ethyl acetate :

$$CH_3COCl + HOC_2H_5 = HCl + CH_3COOC_2H_5.$$
Acetyl chloride.   Ethyl alcohol.   Ethyl acetate.

Acetic Anhydride, $(CH_3CO)_2O$, acetyl oxide, is obtained by the action of acetyl chloride upon dry sodium acetate :

$$CH_3COONa + CH_3COCl = (CH_3CO)_2O + NaCl.$$
Sodium acetate.   Acetyl chloride.   Acetic anhydride.

It is a colourless liquid which boils at 136° C. It is heavier than water, sinking to the bottom as an oil, but reacts gradually with it forming acetic acid:

$$(CH_3CO)_2O + H_2O = 2CH_3COOH.$$
Acetic anhydride.   Acetic acid.

This change takes place immediately with hot water.

Acetic anhydride acts similarly to acetyl chloride, but less energetically, upon substances which contain the hydroxyl group. Both reagents replace the hydrogen of the hydroxyl by an acetyl group, CH_3CO, and both are much used for the purpose of determining the number of hydroxyl groups, if any, present in the molecule of any compound.

Both reagents act also upon the SH group of mercaptans, and upon the NH_2 or NH group of primary or secondary amines. They do not act at all readily upon the hydroxyl group in organic acids.

The number of acetyl groups introduced in place of hydrogen atoms by the action of acetyl chloride or acetic anhydride can generally be discovered by analysis (combustion) of the body formed. In some cases it is better to proceed by expelling the acetyl groups from their combination by boiling the substance with a measured volume of standard alkali solution, and determining the amount of alkali left in excess of what was needed to neutralise the liberated acetic acid. The method may be exemplified by a simpler case than any to which it would in practice be applied:

$$CH_3COOC_2H_5 + NaOH = CH_3COONa + C_2H_5OH.$$
Ethyl acetate.   Sodium acetate.   Ethyl alcohol.

Each molecule of alkali spent in neutralising liberated acetic acid represents one acetyl group in the body examined.

EXPT. 17. In a small strong bottle place 25 c.c. of normal sodium hydrate solution (40 grams NaOH to the litre), and then run in 1 c.c. of ethyl acetate from a 1 c.c. pipette. Fit the bottle immediately with an indiarubber stopper, firmly tied down. Put the bottle in a beaker of water, and heat the water to boiling. Allow to cool. Then wash out the contents of the bottle, and determine how much normal sulphuric acid (49 grams H_2SO_4 to the litre) is required to neutralise the sodium hydrate left uncombined.

Questions on Chapter X

1. Describe how you would prepare acetic anhydride.
2. Explain the use of acetyl chloride and acetic anhydride in investigating the constitution of an organic compound.
3. Glycerine, $C_3H_5(OH)_3$, forms a derivative, $C_3H_5(OC_2H_3O)_3$, when treated with acetic anhydride. 1.026 gram of this glycerine acetate is heated with 25 c.c. of normal sodium hydrate until complete decomposition into glycerine and acetic acid is effected. How many c.c. of normal sulphuric acid would be required to neutralise the sodium hydrate left in excess?

CHAPTER XI

THE AMINES

THE amines have for their inorganic type ammonia, NH_3, and are derived from it by replacement of the hydrogen atoms with alkyl groups. If only one of the three hydrogens is replaced we have a *primary* amine, such as methylamine, NH_2CH_3. When two alkyl groups are introduced we have a *secondary* amine, such as dimethylamine, $NH(CH_3)_2$; while in a *tertiary* amine all three hydrogens are replaced as in trimethylamine, $N(CH_3)_3$.

The power possessed by ammonia of combining with acids to form neutral bodies—the ammonium salts—is retained by its alkyl derivatives. Indeed, the amines with which we are now concerned, and in which methyl and ethyl groups take the place of the hydrogen of the ammonia, are more strongly basic than ammonia itself, while resembling it very markedly in general chemical behaviour. The lower members of the series of amines are gases or volatile liquids smelling strongly of ammonia, but differing from it in being inflammable. They combine directly with acids to form salts, such as methylamine hydrochloride, $NH_2(CH_3) \cdot HCl$, which resemble the ammonium salts in many respects, but differ from them in being soluble in alcohol.

Very important are the double salts which the hydrochlorides of the amines form with platinum tetrachloride. These correspond exactly to the ammonium compound, $(NH_4Cl)_2 \cdot PtCl_4$, frequently used as a means of detecting and estimating ammonia. Like that body, they are decomposed on ignition, leaving a residue of metallic platinum, and in this way it is easy to determine the percentage of that metal in any of these double salts. On examining the formulæ of these compounds:

$$(NH_3 . HCl)_2 PtCl_4, \qquad (NH_2CH_3 . HCl)_2 PtCl_4,$$

we see that each salt contains two molecules of ammonia, or an amine, for each atom of platinum, and we may write the formula as $M_2H_2PtCl_6$, where M represents the amine. The atomic weight of platinum is 198, and if therefore we calculate from the percentage of platinum found by ignition how many parts by weight of the double salt contain 198 parts of that metal, we have the molecular weight of the salt. On subtracting from this 413 $(2 + 198 + 213 : H_2PtCl_6)$, the difference is twice the molecular weight of the amine. This is an easy and accurate method of determining that constant, often the best way of identifying an unknown member of the amine series.

Example. The platinum double salt of an amine gave the following results on ignition: .1935 gram yielded .077 gram platinum. Calculate the molecular weight of the amine.

These numbers show that 198 parts of platinum are contained in $198 \times \frac{1935}{770} = 483.3$ parts of the double salt. Therefore

$$M = \tfrac{1}{2}(483.3 - 413),$$
$$= 35.1,$$

and the amine is either dimethylamine, $NH(CH_3)_2$, or ethylamine, $NH_2C_2H_5$, both of which have the molecular weight 35. To distinguish between these two isomers would require further experiment. See p. 84.

General Method of Preparation of the Amines.—The chief method was discovered by Hofmann, and consists in heating together ammonia (in alcoholic solution) and an alkyl iodide or bromide. By this reaction, which requires a temperature of about 100° C. and the use of sealed glass tubes such as are employed in Carius's method of analysis (see p. 9), a mixture of the iodides of primary, secondary, and tertiary amines is obtained, and the following equations represent the changes which occur:

(1) $NH_3 + CH_3I = NH_2CH_3 . HI.$
Methyl-ammonium iodide.

(2) $NH_3 + 2CH_3I = NH(CH_3)_2 . HI + HI.$
Dimethyl-ammonium iodide.

(3) $NH_3 + 3CH_3I = N(CH_3)_3 . HI + 2HI.$
Trimethyl-ammonium iodide.

At the same time there is produced some amount of a compound, $N(CH_3)_4I$, belonging to the class of quaternary ammonium salts:

(4) $\quad NH_3 + 4CH_3I \quad = \quad N(CH_3)_4I + 3HI.$
$\qquad\qquad\qquad\qquad\qquad$ Tetra-methyl-ammonium iodide.

The proportions in which these four substances are produced depend on the particular alkyl iodide employed. In any case their separation is a matter of considerable labour and difficulty.

The quaternary ammonium salt (such as $N(CH_3)_4I$) is not decomposed on boiling with soda or potash, and is thus easily separated from the three amines, of which a mixture is liberated on treating with alkali the product obtained by Hofmann's method. The tertiary amine, $N(CH_3)_3$, is alone left unacted upon when this mixture is treated with nitrous acid, while the secondary amine is converted into a nitrosamine:

$\quad NH(CH_3)_2 + HNO_2 \quad = \quad (CH_3)_2N.NO \quad + \quad H_2O,$
$\quad$ Dimethylamine. $\qquad\qquad\quad$ Dimethyl-nitrosamine.

and the primary amine into an alcohol:

$\quad NH_2CH_3 + HNO_2 \quad = \quad CH_3.OH \quad + \quad N_2 + H_2O.$
$\quad$ Methylamine. $\qquad\qquad\quad$ Methyl alcohol.

It is not very difficult thus to isolate the tertiary amine, and from the nitrosamine the secondary base can be recovered. The primary amine is, however, lost, but there are other much more convenient ways for the preparation of the primary amines, so that this disadvantage is not very serious.

The Primary Amines of the type $R.NH_2$, where R represents an alkyl group (CH_3, C_2H_5, etc.), are more strongly basic than ammonia itself. They are produced in Hofmann's general reaction, but are difficult to separate from the resulting mixture of ammonium salts. Several methods are known by which primary amines can be obtained in a state of purity, and the most important of these is also due to Hofmann. The amide (see p. 88) of an organic acid, on treatment with bromine and potash, undergoes a peculiar kind of partial oxidation:

$\quad RCONH_2 + O = R.NH_2 + CO_2,$
$\qquad$ Amide. $\qquad\qquad$ Amine.

and the primary amine containing one carbon atom less in the molecule is obtained perfectly free from secondary or tertiary base.

In reality an intermediate product $R . CONHBr$ is first formed:

$$R . CONH_2 + Br_2 + KOH = RCONHBr + KBr + H_2O,$$

and is then decomposed by more potash:

$$R . CONHBr + 3KOH = RNH_2 + K_2CO_3 + KBr + H_2O.$$

The primary amines react very readily with many reagents. We can only refer here to the formation of alcohols from them by the action of nitrous acid:

$$\underset{\text{Primary amine.}}{RNH_2} + HNO_2 = \underset{\text{Alcohol.}}{R . OH} + N_2 + H_2O.$$

The Secondary Amines, R_2NH, are best distinguished from primary and tertiary amines by the formation of nitrosamines, $R_2N . NO$, on treatment with nitrous acid:

$$\underset{\text{Secondary amine.}}{R_2NH} + HNO_2 = \underset{\text{Nitrosamine.}}{R_2N . NO} + H_2O.$$

These nitrosamines are mostly oils, insoluble in water, which regenerate the amine when boiled with concentrated hydrochloric acid:

$$R_2N . NO + HCl = R_2NH + NOCl.$$

The Tertiary Amines, R_3N, are still more strongly basic than either primary or secondary amines containing the same alkyl groups. They are unacted upon by many reagents which readily affect the other two classes of amines, such as nitrous acid, acetic anhydride, etc., and can thus be easily distinguished or separated from them.

Methylamine, NH_2CH_3, is a colourless gas, very soluble in water, and possessing a strong smell of ammonia. It is combustible. The most convenient method of preparing it in the laboratory is by the action of bromine and potash on acetamide, or a reaction represented by the following equation:

$$\underset{\text{Acetamide.}}{CH_3CONH_2} + KOH + Br_2 = \underset{\text{Methylamine.}}{CH_3NH_2} + 2KBr + K_2CO_3 + 2H_2O.$$

A substance, $CH_3CONHBr$, is formed as an intermediate product. For experimental details a book on Organic Preparations should be consulted.

Methylamine is strongly alkaline, and combines readily with acids. Its hydrochloride, $NH_2CH_3 . HCl$, is very soluble in water. The platinum double salt has the formula $(NH_2CH_3 . HCl)_2 PtCl_4$, and is only slightly soluble in water, from which it crystallises in minute hexagonal plates, whose appearance under the microscope is very characteristic.

Dimethylamine, $NH(CH_3)_2$, is a volatile liquid boiling at 7°. Its odour is at once somewhat fishy and strongly ammoniacal. It is present in herring-brine, and in large quantity in a by-product of the manufacture of beet-sugar.

Dimethylamine is most conveniently made on a small scale from a derivative of aniline. This substance, which is to be described more fully in Part II., is a primary amine of the formula $C_6H_5 . NH_2$, and can, by treatment with methyl chloride, be converted into $C_6H_5 . N(CH_3)_2$, dimethylaniline. From this, by a peculiar reaction which we cannot here describe further, it is possible to prepare pure dimethylamine.

Trimethylamine, $N(CH_3)_3$, in the undiluted state smells of ammonia, but when largely mixed with air (possibly owing to some kind of oxidation) has a strong odour of rotten fish. It is present in herring-brine, as also in the by-product, already referred to, of the beet-sugar manufacture. This product is obtained by the dry distillation of beet-root molasses, and contains all three methylamines but dimethylamine in largest quantity. This mixture has been largely used in the manufacture of methyl chloride, as all the methylamines when heated with concentrated HCl are converted into ammonia with separation of methyl chloride:

$$N(CH_3)_3 . HCl + 2HCl = NH_3 + 3CH_3Cl,$$
$$NH(CH_3)_2 . HCl + HCl = NH_3 + 2CH_3Cl,$$
$$NH_2CH_3 . HCl = NH_3 + CH_3Cl.$$

Tetramethyl-ammonium Iodide, $N(CH_3)_4 I$, is the chief product of the action of NH_3 on CH_3I. The radicle $N(CH_3)_4$ is so strongly electro-positive that the compound $N(CH_3)_4 I$ is not decomposed by potash or soda, but on treating it with moist silver oxide, the ammonium hydrate $N(CH_3)_4OH$ is obtained:

$$2N(CH_3)_4I + Ag_2O + H_2O = 2N(CH_3)_4OH + 2AgI.$$

The solution of this hydrate is as strongly alkaline as a solution of potash or soda, and it is interesting to notice the more and more marked alkaline character of the compounds

$$NH_3,\ NH_2CH_3,\ NH(CH_3)_2,\ N(CH_3)_3,\ N(CH_3)_4OH,$$

as the number of alkyl groups attached to the nitrogen atom is increased.

Ethylamine, $NH_2C_2H_5$, is obtained along with di- and tri-ethylamines by heating together in sealed tubes ethyl iodide and alcoholic ammonia. Like other primary amines, it is decomposed by nitrous acid:

$$NH_2C_2H_5 + HNO_2 = C_2H_5OH + N_2 + H_2O,$$

whereas **Diethylamine,** $NH(C_2H_5)_2$, forms a nitrosamine, $(C_2H_5)_2N \cdot NO$, and **Triethylamine** is unaltered.

QUESTIONS ON CHAPTER XI

1. What substances are formed when ethyl iodide is heated in sealed tubes with an alcoholic solution of ammonia?

2. How can you distinguish the three classes of amines, primary, secondary, and tertiary?

3. How would you prepare mono-ethylamine? What is the action upon it of hydrochloric and of nitrous acids.

4. The platinum double salt of an amine gave the following result upon analysis: .2055 gram of the double salt left .0680 gram of platinum upon ignition. Calculate the molecular weight of the amine.

CHAPTER XII

THE AMIDES AND AMIDO-ACIDS

The Amides, like the amines, are derived from ammonia; but whereas in the amines the hydrogen of the NH_3 is replaced by an alcohol radicle (or "alkyl" group), such as methyl, CH_3, or ethyl, C_2H_5, in the amides the substituting group is an acid radicle.

These acid radicles are those which yield acids when combined with the hydroxyl group OH (see p. 68), the radicle of acetic acid being $CH_3.CO$, acetyl, that of propionic acid being $C_2H_5.CO$, propionyl. *Acet*-amide is the name given to the amide CH_3CONH_2, in which one of the three hydrogens in the ammonia has been replaced by the acetyl group, whilst *propion*-amide is the name used for $C_2H_5CONH_2$.

Methods of Formation.—(1) The first of these is by the action of ammonia upon the chloride of the acid radicle; *e.g.* acetyl chloride yields acetamide:

$$CH_3COCl + 2NH_3 = CH_3CONH_2 + NH_4Cl.$$
Acetyl chloride. Acetamide.

This is analogous to the preparation of the amines from alkyl iodides (or chlorides) and ammonia; but the reaction occurs far more readily with the acid chlorides, so that it is often necessary to employ means to moderate the violence of the action.

(2) The second method only differs from the first in the use of an ethereal salt of the acid instead of the acid chloride.

Thus ethyl acetate when heated with ammonia gives acetamide and ethyl alcohol:

$$CH_3CO|OEt + H|NH_2 = CH_3CONH_2 + EtOH.$$

Ethyl acetate.   Acetamide.

This method in many cases requires the action of a high temperature (150° C.) for several hours, and involves therefore the use of closed vessels which can stand very considerable pressure. On the small scale, sealed glass tubes similar to those used in Carius's method of analysis (p. 9) are employed, but unless very well made, these cannot stand the pressure of the strong aqueous ammonia at 150° C.

(3) The third method consists in heating strongly (230° C.) the ammonium salt of the acid, when it loses water and is converted into the amide:

$$CH_3COONH_4 = CH_3CONH_2 + H_2O.$$

Ammonium acetate.   Acetamide.

Here again it is generally necessary to heat in closed vessels (sealed glass tubes), but the pressures produced are not great, and there is little risk of the tubes being burst.

Acetamide, CH_3CONH_2, is best prepared by the third method. It is separated from unaltered ammonium acetate by distillation, and is obtained as a white solid, usually possessing a characteristic odour of mice, but apparently odourless when quite pure. It is not markedly acid or basic in behaviour, the basic character of the ammonia being removed by the introduction of the acid radicle.

When boiled with much water acetamide is slowly converted into ammonium acetate:

$$CH_3CONH_2 + H_2O = CH_3COONH_4,$$

Acetamide.   Ammonium acetate.

and the same change takes place much more readily when an acid or alkali is present.

With an alkali ammonia is evolved; with an acid the ammonium salt of that acid is formed.

The amido-acids are derived from acids such as acetic, $CH_3 . CO_2H$, by introduction of an amido-group, NH_2, in place of one of the hydrogens of the alkyl radicle (in this case methyl, CH_3); thus amido-acetic acid is $NH_2CH_2 . CO_2H$.

The amido-acids are neutral to litmus, but chemically act both as acids and bases. They form salts with copper, silver, and other metals on the one hand, while on the other they combine with strong acids to form salts, such as $HCl.NH_2CH_2CO_2H$, in which the amido-acid plays the part of a substituted ammonia.

Many amido-acids occur in various animal tissues, and may be prepared from them. The most important and instructive artificial methods for their preparation are the following:—

(1) By acting with ammonia on a halogen derivative of the acid, *e.g.* monochlor-acetic acid, $CH_2Cl . CO_2H$, yields amido-acetic acid, $NH_2CH_2 . CO_2H$:

$$CH_2Cl . CO_2H + 2\ NH_3 = NH_2CH_2CO_2H + NH_4Cl.$$
Monochlor-acetic acid. Amido-acetic acid.

In practice the ammonia is obtained from solid ammonium carbonate, which is heated with the monochlor-acetic acid.

Monochlor-acetic acid is formed when chlorine acts upon acetic acid:

$$CH_3CO_2H + Cl_2 = CH_2Cl . CO_2H + HCl.$$

(2) The second method, somewhat more complicated, starts from the aldehydes. These are able to combine with ammonium cyanide to form compounds, such as $CH_3 . CH(NH_2)(CN)$:

$$CH_3 . CHO + NH_4CN = CH_3 . CH{<}{}^{NH_2}_{CN} + H_2O.$$

These when treated with dilute acids (see p. 69) yield amido-acids by conversion of the CN group into carboxyl, CO_2H.

Amido-Acetic Acid, $NH_2 . CH_2CO_2H$ (**Glycocoll**), can be extracted from glue (therefore from bones) by treatment with sulphuric acid; it is best prepared synthetically from monochlor-acetic acid by the action of ammonia. It forms a dark-blue copper salt $(NH_2CH_2CO_2)_2Cu$, and also a chloride,

HCl . $NH_2CH_2CO_2H$. Glycocoll itself is a white crystalline solid, having a sweetish taste, and readily soluble in water.

Of the more complicated amido-acids we may mention **Leucin**, $C_4H_9 . CHNH_2 . CO_2H$, therefore **amido-caproic acid**, which is one of the most important products of the decomposition of albumen, either by putrefaction or by boiling with acids or alkalies; it is also present ready-formed in various glands of the body as the pancreas, spleen, etc.

QUESTIONS ON CHAPTER XII

1. What substances are formed by the action of ammonia upon (*a*) monochlor-acetic acid, (*b*) methyl chloride, (*c*) acetyl chloride? Give equations.

2. What is the action of dilute hydrochloric acid upon (*a*) acetamide, (*b*) glycocoll?

3. Describe the preparation of acetamide and of glycocoll from acetic acid.

CHAPTER XIII

ALKYL COMPOUNDS OF PHOSPHORUS, ARSENIC, SILICON, AND THE METALS

COMPOUNDS OF PHOSPHORUS

Phosphines.—Phosphine, PH_3, is much less basic in character than ammonia, but is yet capable of combining with hydrogen iodide to form the compound phosphonium iodide, PH_4I. The organic phosphines, obtained by substituting the hydrogen in PH_3 by alkyl groups, are stronger bases in proportion to the number of alkyl groups introduced.

By the action of alkyl iodides on PH_3 only tertiary phosphines, PR_3, and phosphonium compounds, PR_4I, are obtained. The primary and secondary bases can, however, be prepared by employing, instead of PH_3, a mixture of PH_4I with oxide of zinc. The separation of these bodies is not then difficult, and depends on the gradual increase in their basic character from primary phosphine to phosphonium compounds:

- Tetra-methyl phosphonium iodide, $P(CH_3)_4I$, is not decomposed by KOH.
- Tri-methyl phosphonium iodide, $P(CH_3)_3 \cdot HI$, is decomposed by KOH.
- Dimethyl phosphonium iodide, $HP(CH_3)_2 \cdot HI$, is not decomposed by water.
- Methyl phosphonium iodide, $H_2PCH_3 \cdot HI$, is decomposed by water.

The phosphines are gases or volatile liquids, very inflammable, and possessed of very strong unpleasant odours. Characteristic of the tertiary phosphines is the readiness with which they combine with O, S, Cl_2, etc., to form compounds such as $P(C_2H_5)_3O$, triethyl-phosphine oxide, in which the radicle $P(C_2H_5)_3$ plays the part of a divalent metal.

Methyl-phosphine, $PH_2(CH_3)$, is obtained along with diethyl-phosphine, $PH(CH_3)_2$, by heating a mixture of phosphonium iodide with methyl iodide and zinc oxide in sealed tubes to a temperature of 150° C. :

$$2PH_4I + 2CH_3I + ZnO = 2PH_2(CH_3) \cdot HI + ZnI_2 + H_2O,$$
Phosphonium iodide.   Methyl-phosphonium iodide.

$$PH_4I + 2CH_3I + ZnO = PH(CH_3)_2 \cdot HI + ZnI_2 + H_2O.$$
Dimethyl-phosphonium iodide.

On heating the resulting product with water, the methyl-phosphine, PH_2CH_3, is liberated, while the dimethyl-phosphine, $PH(CH_3)_2$, remains combined with the hydriodic acid. On subsequently boiling with sodium hydrate, the secondary phosphine is in turn set free.

Trimethyl-phosphine, $P(CH_3)_3$, is prepared by heating phosphonium iodide with methyl iodide without the addition of zinc oxide :

$$PH_4I + 3CH_3I = P(CH_3)_3 \cdot HI + 3HI.$$
Trimethyl-phosphonium iodide.

The phosphine is obtained when the product is decomposed by boiling with potash or soda.

All three of these methyl-phosphines are very inflammable, and exhibit intolerable odours ; they readily combine with chlorine, bromine, oxygen, or sulphur to form compounds such as $P(CH_3)_3O$ or $PH(CH_3)_2Cl_2$.

COMPOUNDS OF ARSENIC

The Arsines are connected with AsH_3 as the amines with NH_3. No primary or secondary arsines (such as AsH_2CH_3) are known, but tertiary compounds, $As(CH_3)_3$, etc., have been prepared. These, like the parent substance AsH_3, are incapable of forming salts.

The most important organic compounds of arsenic form a series in which the radicle $As(CH_3)_2$ corresponds to an atom of a monovalent metal, and it has been found convenient to give a special name—**cacodyl**—to this radicle.

Cacodyl Oxide, $\{As(CH_3)_2\}_2O$ or Cd_2O, can be obtained by distilling As_2O_3 with potassium acetate:

$$4CH_3 . CO_2K + As_2O_3 = \begin{matrix} As(CH_3)_2 \\ As(CH_3)_2 \end{matrix}\!\!>\!O + 2K_2CO_3 + 2CO_2,$$

Potassium acetate.   Cacodyl oxide.

and from this oxide, by the action of acids, various salts can be made, amongst them **cacodyl chloride**, $As(CH_3)_2Cl$, which when reduced with zinc furnishes free **cacodyl**, $As_2(CH_3)_4$. All of these compounds are liquids of disgusting odour and intensely poisonous properties. They are also very inflammable, and their investigation, which was carried out by Bunsen, was a matter of great danger and difficulty.

Of the organic compounds of arsenic, other than the cacodyl derivatives, we will mention only one.

Arsenic Trimethyl, $As(CH_3)_3$, which is obtained by the action of methyl iodide upon an alloy of arsenic and sodium. It is a colourless volatile liquid of unpleasant smell, readily combining with one atom of oxygen or two of chlorine to form compounds in which the arsenic is pentavalent, e.g. $As(CH_3)_3O$, $As(CH_3)_3Cl_2$. It is not basic in character, and has no tendency to combine with acids to form salts in the same way as $N(CH_3)_3$ and $P(CH_3)_3$.

COMPOUNDS OF SILICON

The organic derivatives of the tetravalent element silicon may be put side by side with other organic compounds in which the place of the silicon is taken by carbon, itself a tetravalent element; thus—

Silicon Tetramethyl, $Si(CH_3)_4$, may be compared with the pentane $C(CH_3)_4$; it is obtained by treating silicon tetrachloride with zinc methyl (see p. 95):

$$SiCl_4 + 2Zn(CH_3)_2 = Si(CH_3)_4 + 2ZnCl_2,$$

and is a volatile liquid unaltered by water. The analogy with the pentane, $C(CH_3)_4$, is not merely one of formulæ, but is also seen to some extent in the chemical behaviour of the two compounds. This is not surprising in view of the fact that both are of similar type, and each contains four methyl groups; but that there exists no complete analogy between the silicon derivatives and those of carbon is evident from the great differences between the simple corresponding derivatives of the two elements; thus silico-chloroform, $SiHCl_3$, is a liquid fuming in the air and decomposed by water, therefore quite unlike chloroform itself, $CHCl_3$.

COMPOUNDS OF THE METALS WITH ALKYL RADICLES

Many of the metals form alkyl compounds, usually volatile liquids which oxidise rapidly or even ignite spontaneously in the air; they are obtained by the action of alkyl iodides upon the metals, or upon their alloys with zinc or sodium.

A second method is to act with zinc methyl (or zinc ethyl, etc.) upon the chloride of the metal which is to be converted into an alkyl derivative.

Zinc Methyl, $Zn(CH_3)_2$, is obtained by the action of methyl iodide upon zinc and subsequent distillation. In the first place, direct combination of the zinc and methyl iodide occurs:

$$Zn \ + \ CH_3I \ = \ Zn{<}^{CH_3}_{I}.$$

Methyl iodide.   Zinc methyl-iodide.

The compound thus formed is decomposed by further heating:

$$2Zn{<}^{CH_3}_{I} \ = \ Zn(CH_3)_2 + ZnI_2.$$

Zinc methyl-iodide.   Zinc methyl.

The first reaction takes place more readily when the zinc-copper couple of Gladstone and Tribe is used instead of zinc filings. This couple is merely an intimate mixture of finely divided copper and zinc, and can be most readily prepared by mixing zinc filings with one-ninth of their weight of copper-dust, obtained by reducing powdered copper oxide in a current of hydrogen. The couple only requires to be heated for a few minutes in a flask to make it ready for use.

EXPT. 18. Mix 90 grams zinc filings with 10 grams of reduced copper. Place the mixture in a flask fitted with cork and capillary tube, and heat for a few minutes over the bare flame of a Bunsen burner. When cool,

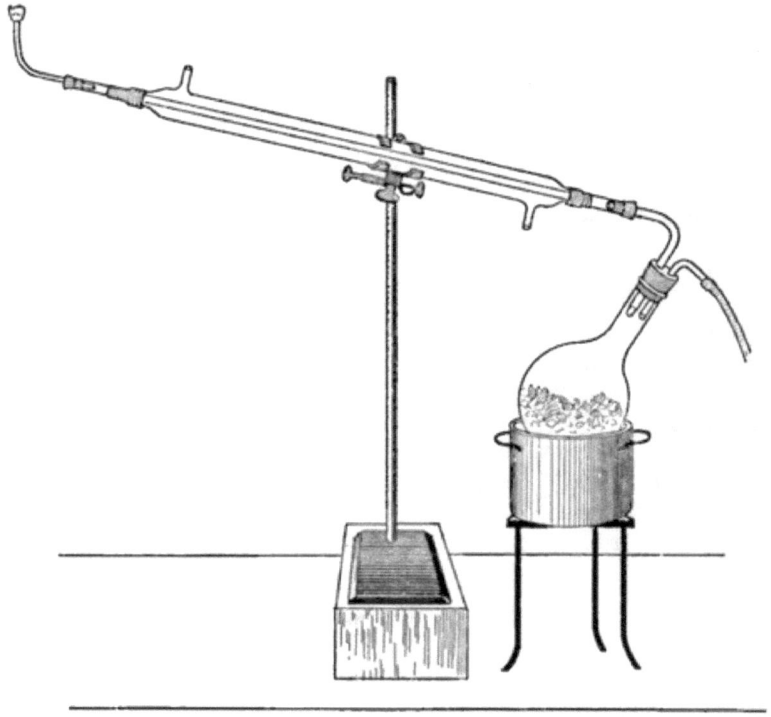

FIG. 29. - Preparation of Zinc Ethyl.

add 50 grams methyl iodide, and fit the flask with an inverted condenser and a tube for introducing coal-gas (see fig. 29). Heat on the water bath for ten hours; then arrange the condenser for distillation and distil off the zinc methyl in a slow current of coal-gas.

Zinc Ethyl, $Zn(C_2H_5)_2$, is similarly prepared. Both are important laboratory reagents for the purpose of introducing methyl or ethyl groups in the place of chlorine or other element, *e.g.* a ketone can be made by the action of zinc ethyl on acetyl chloride :

$$2CH_3 . CO . Cl + Zn(C_2H_5)_2 = 2CH_3 . CO . C_2H_5 + ZnCl_2.$$
Acetyl chloride.   Methyl-ethyl ketone.

Zinc methyl and ethyl fume in the air and very readily take fire, often spontaneously. They are decomposed by water :

$$Zn(CH_3)_2 + 2H_2O = Zn(OH)_2 + 2CH_4,$$
Zinc methyl.   Methane.

and by the halogens :

$$Zn(C_2H_5)_2 + 2Br_2 = ZnBr_2 + 2C_2H_5Br.$$
Zinc ethyl.   Ethyl iodide.

Mercury Methyl, $Hg(CH_3)_2$, and **Mercury Ethyl**, $Hg(C_2H_5)_2$, are colourless liquids, whose vapours have only feeble odour, but are very poisonous ; they are prepared from sodium amalgam by the action of methyl or ethyl iodide :

$$NaHg + 2CH_3I = 2NaI + Hg(CH_3)_2.$$
Methyl iodide.   Mercury methyl.

They are more stable than the zinc compounds, and neither take fire in the air nor are decomposed by water. Their general chemical behaviour is otherwise similar to that of their zinc analogues.

Alkyl Compounds of other Metals. — Many other metals also form similar compounds, the most important cases being perhaps those of lead and tin.

By acting on lead chloride, $PbCl_2$, with zinc ethyl the substance **lead tetra-ethyl**, $Pb(C_2H_5)_4$, is obtained :

$$2PbCl_2 + 2Zn(C_2H_5)_2 = Pb(C_2H_5)_4 + Pb + 2ZnCl_2,$$

a fact which proves that lead is really a tetravalent element, and is thus in complete agreement with the recent discovery of the existence of a lead tetrachloride, $PbCl_4$. Lead tetra-ethyl is an oily liquid which takes fire when heated in contact with air.

In the same way, by acting on stannous chloride, $SnCl_2$, with zinc ethyl, we obtain **tin tetra-ethyl**, $Sn(C_2H_5)_4$, in which the maximum valency of the metal is exerted:

$$2SnCl_2 + 2Zn(C_2H_5)_2 = Sn(C_2H_5)_4 + Sn + 2ZnCl_2;$$
$$\text{Zinc ethyl.} \qquad \text{Tin tetra-ethyl.}$$

but **tin di-ethyl**, $Sn(C_2H_5)_2$, can be got by treating an alloy of tin and sodium with ethyl iodide:

$$SnNa + 2C_2H_5I = Sn(C_2H_5)_2 + 2NaI.$$
$$\text{Ethyl iodide.} \quad \text{Tin di-ethyl.}$$

$Sn(C_2H_5)_4$ is a liquid which can be distilled without decomposition, whereas $Sn(C_2H_5)_2$ decomposes into the tetra-ethyl compound and metallic tin.

QUESTIONS ON CHAPTER XIII

1. Give the preparation of (*a*) phosphonium iodide, (*b*) tetra-methyl phosphonium iodide. What is the action of water and of potassium hydrate solution upon these two compounds?

2. How is free cacodyl obtained? Give the formulæ of cacodyl oxide and cacodyl chloride.

3. Give the preparation of zinc ethyl. What is the action upon it of (*a*) water, (*b*) chlorine, (*c*) acetyl chloride?

4. What reasons have we for considering lead to be a tetra-valent metal?

CHAPTER XIV

GLYCOL AND ITS DERIVATIVES. SUCCINIC, MALIC, AND TARTARIC ACIDS

Glycol, $C_2H_4(OH)_2$, is a substance containing two hydroxyls combined with the divalent radicle, C_2H_4, ethylene. Each of these hydroxyls behaves similarly to the hydroxyl in an alcohol, so that glycol may be termed a dihydric alcohol.

It is obtained from ethylene bromide, $C_2H_4Br_2$, by replacement of the bromine:

$$\underset{\text{Ethylene bromide.}}{C_2H_4Br_2} + 2HOH = \underset{\text{Glycol.}}{C_2H_4(OH)_2} + 2HBr,$$

just as the ethyl alcohol can be got from ethyl bromide:

$$C_2H_5Br + HOH = C_2H_5OH + HBr.$$

It is necessary when preparing glycol from ethylene bromide in this way to heat with a large quantity of water to temperatures of 150° C. or thereabouts; the reaction takes place more readily when sodium carbonate is added to the water.

This method of preparation indicates the constitutional formula $\begin{array}{l} CH_2OH \\ | \\ CH_2OH \end{array}$ for glycol, according to which it may be regarded as a *dihydric primary alcohol*, and this view is strengthened by consideration of the substance's general chemical behaviour.

As a *dihydric alcohol* glycol reacts with sodium or potassium to form glycolates analogous to the alcoholates (p. 44):

$$C_2H_4(OH)_2 + Na = C_2H_4{<}^{OH}_{ONa} + H,$$

and

$$C_2H_4(OH)_2 + 2Na = C_2H_4(ONa)_2 + H_2;$$

and ethereal salts of glycol can be obtained by the action on it of acids:

$$\underset{\text{Glycol.}}{C_2H_4(OH)_2} + \underset{\text{Acetic acid.}}{2CH_3CO_2H} = \underset{\text{Ethylene acetate.}}{C_2H_4(OCO_2CH_3)_2} + 2H_2O;$$

just as ethyl acetate is got from ethyl alcohol and acetic acid, so here ethylene acetate is obtained.

As a dihydric *primary* alcohol glycol furnishes, when treated with oxidising agents, bodies in which the groups CH_2OH are successively oxidised to aldehyde groups CHO, and finally to carboxyl groups CO_2H. The substances thus obtained are presently to be considered.

Glycol is a thick colourless liquid with a sweetish taste.

<small>The isomeric glycol, ethylidene glycol, $CH_3.CH(OH)_2$, does not seem able to exist unless in dilute solution; instead of this we obtain aldehyde $CH_3.CHO$ when the ethylidene glycol might be expected, *e.g.* in action of water on $CH_3.CHCl_2$.</small>

Glyoxal, $(CHO)_2$, is the di-aldehyde of glycol, and is formed along with other substances in the oxidation of glycol:

$$\underset{\text{Glycol.}}{\begin{array}{c}CH_2OH\\|\\CH_2OH\end{array}} + O = \underset{\text{Glyoxal.}}{\begin{array}{c}CHO\\|\\CHO\end{array}} + H_2O.$$

Like other aldehydes, it readily reduces Fehling's solution or ammoniacal silver solution (see p. 62).

Glycolic Acid, $CH_2(OH).CO_2H$, is also formed in the oxidation of glycol:

$$\underset{\text{Glycol.}}{\begin{array}{c}CH_2OH\\|\\CH_2OH\end{array}} + O_2 = \underset{\text{Glycolic acid.}}{\begin{array}{c}CH_2OH\\|\\CO_2H\end{array}} + H_2O,$$

but is better prepared from mono-chloracetic acid by boiling with water, to which calcium carbonate in fine powder has been added (to combine with the HCl):

$$CH_2Cl \cdot CO_2H + H_2O = CH_2(OH) \cdot CO_2H + HCl.$$
Monochlor-acetic acid. ⸺⸺⸺⸺⸺ Glycolic acid.

Glycolic acid forms white crystals. As an acid it yields well-defined salts, such as silver glycolate, $CH_2OH \cdot CO_2Ag$, while as an alcohol it combines with acids to produce ethereal salts.

Oxalic Acid, $(CO_2H)_2$, is a more completely oxidised product of glycol:

$$\begin{array}{c} CH_2 \cdot OH \\ | \\ CH_2 \cdot OH \end{array} + 2O_2 = \begin{array}{c} CO_2H \\ | \\ CO_2H \end{array} + 2H_2O.$$
Glycol. ⸺⸺⸺⸺ Oxalic acid.

It is an important acid, the starting-point of a series of organic dibasic acids. Oxalic acid is found in many plants, especially the varieties of *Oxalis*, and can be prepared artificially in several ways, of which three further ones (in addition to the oxidation of glycol) may be mentioned:

(1) Carbon dioxide when passed over heated metallic sodium combines with it to form sodium oxalate:

$$2CO_2 + 2Na = C_2O_4Na_2.$$
Sodium oxalate.

(2) Sodium formate when strongly heated evolves hydrogen and yields sodium oxalate:

$$2H \cdot CO_2Na = H_2 + C_2O_4Na_2$$
Sodium formate. ⸺⸺⸺⸺ Sodium oxalate.

(3) An important practical method for the manufacture of oxalic acid is the action of caustic alkalies upon cellulose. Sawdust (the form of cellulose generally used) is mixed into a paste with a strong solution of potash, and then heated on iron plates. The product is extracted with water, and the oxalic acid separated by precipitation as calcium oxalate.

Oxalic acid forms crystals which contain two molecules of

water of crystallisation. When heated the crystals lose water, and then decompose into formic acid and carbon dioxide:

$$C_2O_4H_2 = CO_2 + H \cdot CO_2H.$$
Oxalic acid.   Formic acid.

Oxalic acid when heated with strong sulphuric acid does not blacken, but is decomposed with evolution of the two oxides of carbon in equal volumes:

$$C_2O_4H_2 = CO + CO_2 + H_2O.$$

Oxalic acid is a stronger acid than acetic, and being a dibasic acid forms two series of stable salts.

Potassium Oxalate, $C_2O_4K_2$, is used in preparing the "ferrous oxalate developer," largely employed in photography.

Potassium Hydrogen Oxalate, C_2O_4KH, along with free oxalic acid, composes the "*salts of lemon*" used for removing ink-stains from cloth.

Ammonium Oxalate, $C_2O_4(NH_4)_2$, is used as a reagent in the laboratory.

SUCCINIC, MALIC, AND TARTARIC ACIDS

Succinic Acid, $C_4H_6O_4$, was first obtained by distillation of amber, and this is still the way prescribed for its preparation in the British Pharmacopœia. It is also present in some other resins and in lignite. The artificial methods for making the acid and its reactions are best represented by the constitutional formula given below; the chief of these methods are:

(1) Ethylene cyanide (from ethylene bromide and AgCN), when boiled with dilute acids or alkalies, yields succinic acid:

$$\begin{array}{c} CH_2 \cdot CN \\ | \\ CH_2 \cdot CN \end{array} + 4H_2O = \begin{array}{c} CH_2 \cdot CO_2H \\ | \\ CH_2 \cdot CO_2H \end{array} + 2NH_3.$$
Ethylene cyanide.   Succinic acid.

(2) Succinic acid is also obtained by the reduction of malic acid, which can itself be similarly obtained from tartaric acid:

$$C_4H_6O_6 - O = C_4H_6O_5.$$
Tartaric acid.   Malic acid.

$$C_4H_6O_5 - O = C_4H_6O_4.$$
Malic acid.   Succinic acid.

The reduction can be effected by heating with hydriodic acid in sealed tubes.

Succinic acid forms colourless crystals, soluble in water, and possessing an unpleasant taste.

Malic Acid, $C_4H_6O_5$, occurs in the juice of apples and of many other fruits. Its close relation to succinic acid is indicated by the reaction, above referred to, by which that acid is obtained by the reduction of malic acid, and the exact character of the relation is made clear by the following method of preparation :—

(1) Malic acid is produced when monobrom-succinic acid is treated with silver oxide and water:

$$\begin{matrix} CHBr.CO_2H \\ | \\ CH_2.CO_2H \end{matrix} + AgOH = \begin{matrix} CH(OH).CO_2H \\ | \\ CH_2.CO_2H \end{matrix} + AgBr.$$

Monobrom-succinic acid.   Malic acid.

Malic acid is therefore monohydroxy-succinic acid.

(2) Malic acid is formed by the partial reduction of tartaric acid:

$$C_6H_4O_6 - O = C_6H_4O_5.$$
Tartaric acid.   Malic acid.

Malic acid forms deliquescent needles. It is a somewhat stronger acid than succinic, and forms several well-crystallised salts. Very important is the existence of three isomeric forms of malic acid which differ chiefly in their action upon polarised light. One form, the ordinary one obtained from berries, rotates the plane of polarisation to the left; a second form, prepared from dextro-tartaric acid, rotates the plane of polarisation to the right; while the third form, obtained synthetically, is inactive. The fuller consideration of this case of isomerism is deferred until Part II. of this book.

Tartaric Acid, $C_4H_6O_6$, is present in the juice of many fruits, especially in that of grapes; practically the only source of the acid is the "argol," an impure potassium tartrate, deposited during the fermentation of grape-juice. The constitutional formula of the acid is evident from its relation to malic and succinic acids (into which it is in turn converted by reduction), and from the following synthetical methods of preparation :—

(1) Dibrom-succinic acid when boiled with water and silver oxide yields tartaric acid :

$$\begin{array}{c} CHBr.CO_2H \\ | \\ CHBr.CO_2H \end{array} + 2AgOH = \begin{array}{c} CH(OH)CO_2H \\ | \\ CH(OH)CO_2H \end{array} + 2AgBr;$$

Dibrom-succinic acid. Tartaric acid.

tartaric acid is accordingly dihydroxy-succinic acid.

Tartaric acid furnishes another instance of the existence of isomers inexplicable by the theory hitherto alone employed for the explanation of cases of isomerism. The isomers again differ, just as was the case with the malic acids, chiefly in their action upon polarised light. Tartaric acid furnishes four such isomers, of which one is dextro-rotatory (rotates the plane of polarisation to the right), another is lævo-rotatory, while the other two are inactive. We shall here consider only the common variety, dextro-tartaric acid, leaving the others to be discussed in Part II.

Dextro-tartaric acid is the tartaric acid of the shops. It is prepared from argol by conversion into calcium tartrate (treatment with milk of lime), and subsequent liberation of the free acid by addition of sulphuric acid. It is purified by recrystallisation, and forms large prismatic crystals which are readily soluble in water. The solution rotates the plane of polarisation of light to the right. It is a dibasic acid, and the following salts formed by it are of importance :—

Potassium Hydrogen Tartrate, $C_4O_6H_5K$, is the "cream of tartar" of the druggist, and is obtained by purifying the "argol" deposited in the fermentation of grape-juice. It is only slightly soluble in water, and hence sodium hydrogen tartrate will precipitate it from solutions of potassium salts, unless very dilute ; this reaction is sometimes used as a test for the presence of potassium in place of the more expensive method by means of platinic chloride.

Potassium Sodium Tartrate, $C_4O_6H_4KNa$, is known as "Rochelle salt," and is prepared by mixing solutions of sodium hydrate and cream of tartar.

Tartar Emetic is the name of a substance which is obtained by boiling cream of tartar and oxide of antimony with water. Its constitution is generally supposed to be represented by the formula $C_4O_6H_4(SbO)K$, according to which one hydrogen atom of the tartaric acid is replaced by the monovalent group $(Sb'''O)$, antimonyl. Tartar emetic is then to be termed potassium antimonyl tartrate.

The same group, SbO, exists in the compound which is obtained as a white precipitate when water is added to a solution of antimony chloride. This precipitate has the composition SbOCl, and is produced according to the equation
$$SbCl_3 + H_2O = SbOCl + 2HCl.$$

Questions on Chapter XIV

1. By what reactions would you proceed to prepare glycol from ethyl alcohol?
2. Show by its reactions that glycol behaves as a dihydric primary alcohol.
3. Give two ways by which oxalic acid can be synthesised from its elements. Describe the commercial process for the manufacture of the acid.
4. What is the relation between succinic, malic, and tartaric acids? How can you pass from each of them to the others?
5. Write down the formulæ of (*a*) salts of lemon, (*b*) tartar emetic, (*c*) cream of tartar, (*d*) succinic acid.

CHAPTER XV

LACTIC AND CITRIC ACIDS

Lactic Acid is a substance present in sour milk which, when isolated and examined as to its chemical relationship, is found to be predominantly an acid, but also to possess some of the properties of alcohols. Its empirical formula is CH_2O as determined by analysis, and as the acid cannot be vaporised without decomposition, we are unable to ascertain its molecular weight by a vapour density determination. It has recently become possible to employ other means for finding the molecular weight of the acid itself, but a little study of the compounds of lactic acid enables us to discover its molecular, and then its constitutional formula.

Lactic acid forms only one sodium salt, sodium lactate, whose analysis indicates the formula $C_3H_5NaO_3$, and therefore the molecular formula $C_3H_6O_3$ for the acid (this agrees with the vapour density of ethyl lactate $C_3H_5O_3 . C_2H_5$). Lactic acid is therefore a monobasic-acid, and contains one carboxyl group, CO_2H. But in this sodium salt there is yet left a hydrogen atom which can with some little difficulty be replaced by sodium, and behaves like the hydrogen atom of an alcoholic hydroxyl. Lactic acid is therefore seen to contain the group OH also.

Lactic acid, $C_3H_6O_3$, may therefore be written $C_2H_4(OH)(CO_2H)$, and the only question left to solve is whether the OH and the CO_2H are connected to the same or to different carbon atoms, whether it is

$$(a)\ \begin{array}{c} CH_2(OH) \\ | \\ CH_2 . CO_2H \end{array} \qquad or\ (b)\ \begin{array}{c} CH_3 \\ | \\ CH{<}^{OH}_{CO_2H} \end{array}$$

Now lactic acid can be got from aldehyde, $CH_3.CHO$, by adding to it HCN, and boiling the product with hydrochloric acid (see pp. 64 and 69):

$$CH_3.CHO \longrightarrow CH_3.CH{<}^{OH}_{CN} \longrightarrow CH_3.CH{<}^{OH}_{CO_2H};$$

Aldehyde.   Lactic acid.

and we are thus led to assign to lactic acid the formula (*b*) of the two given above.

The lactic acid in sour milk is produced from the lactose or milk-sugar present in milk by the action of a particular ferment. Cane-sugar, starch, and other carbohydrates also yield lactic acid under the influence of the same ferment:

$$\underset{\text{Milk or cane sugar.}}{C_{12}H_{22}O_{11}} + H_2O = \underset{\text{Lactic acid.}}{4 C_3H_6O_3}.$$

In preparing lactic acid the following is a good method of procedure:

One kilogram of cane-sugar is dissolved along with about 5 grams of tartaric acid in $3\frac{1}{2}$ litres of water; after a few days some rotten cheese (30 grams) is rubbed into a paste with sour milk ($1\frac{1}{2}$ litres), and added to the solution with 400 grams of zinc oxide. The whole is left to ferment in a warm place for a week or ten days. Then the mixture is heated to boiling, filtered, and the filtrate evaporated. Crystals of zinc lactate separate out on cooling; they are collected and dissolved in water, and the zinc removed by passing H_2S. The zinc sulphide is removed by filtration, and the solution of lactic acid evaporated on the water bath.

Lactic acid thus obtained is a thick syrupy liquid. The sodium salt has the formula $C_3H_5NaO_3$, but when this is heated with metallic sodium, a second atom of the metal is introduced in place of the alcoholic hydrogen, and a substance of the formula $C_3H_4Na_2O_3$ is obtained.

Lactic acid can also be prepared by several synthetical methods:
(1) From aldehyde $CH_3.CHO$ (see above).
(2) From the bromopropionic acid, $CH_3.CH_2Br.CO_2H$, and potash.

Great interest attaches to the existence of an isomeric **paralactic acid** which is present in the juice of meat. This behaves exactly like ordinary lactic acid in nearly every other respect, but differs from it in being able to rotate the plane of polarisation of light. This is connected by Van't Hoff, with the fact that one carbon atom in lactic acid is "asymmetric," that is, connected to four dissimilar radicles. For a fuller account of this theory, see the second part of this book.

There is also known another acid, *hydracrylic*, which is isomeric with lactic acid. The formula (*a*) given above (p. 106) is indicated for it by its formation from ethylene as indicated below :

$$\begin{array}{cccc} CH_2 & CH_2OH & CH_2OH & CH_2OH \\ \parallel \longrightarrow & | \longrightarrow & | \longrightarrow & | \\ CH_2 & CH_2Cl & CH_2 . CN & CH_2 . CO_2H. \\ & (+HClO). & (\text{Action of KCN}). & (\text{Boiling with HCl}). \end{array}$$

Citric Acid is found in lemons, currants, cranberries, and many other sour fruits. It is prepared commercially from lemon or lime-juice by means of the calcium salt.

Its formula is found by analysis to be $C_6H_8O_7$, and it behaves as a tribasic acid. It contains, therefore, three carboxyl groups, and forms salts, such as $C_6H_5O_7K_3$, and ethereal salts, such as $C_6H_5O_7(C_2H_5)_3$. In these the action of acetyl chloride proves the existence of an hydroxyl group (see p. 79). The acid therefore contains one OH and three CO_2H group, and its formula may be written $C_3H_4(OH)(CO_2H)_3$.

Citric acid crystallises in large prisms. As a tribasic acid it forms three series of salts, the three potassium salts being $C_6H_7O_7K$, $C_6H_6O_7K_2$, and $C_6H_5O_7K_3$.

Calcium citrate, $(C_6H_5O_7)_2Ca_3$, is remarkable as being less soluble in hot water than in cold, a property made use of in testing for citric acid.

EXPT. 19. To some solution of citric acid in a test tube add lime water until the reaction is slightly alkaline. No precipitate is formed in the cold, but a white precipitate of calcium citrate appears on boiling.

QUESTIONS ON CHAPTER XV

1. How can lactic acid be obtained from sugar? Why is its formula written $C_3H_6O_3$ and not CH_2O ?

2. Mention some other substances which have the same percentage composition as lactic acid. How could you distinguish them?

3. What happens when (*a*) milk turns sour, (*b*) butter turns rancid, (*c*) wine goes sour?

4. Write down the formulæ of (*a*) the three potassium citrates, (*b*) zinc lactate.

CHAPTER XVI

THE ALLYL COMPOUNDS

THE allyl compounds may be regarded as being derived from the hydrocarbon propylene, C_3H_6, and their starting-point—allyl alcohol—stands to propylene in the same relation as ethyl alcohol does to ethane.

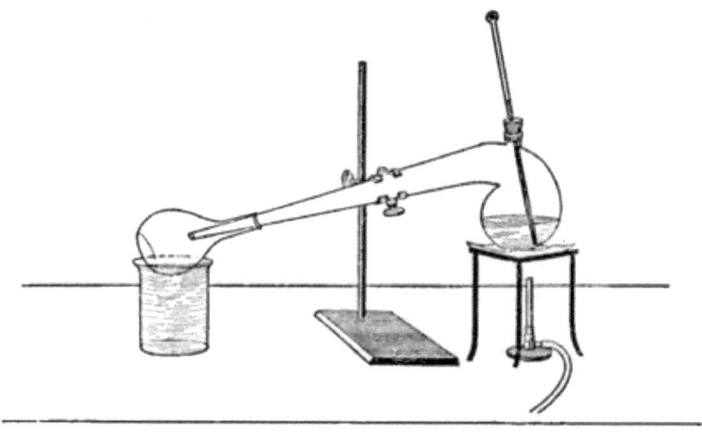

FIG. 30.—Preparation of Allyl Alcohol.

Propylene has the formula $CH_2 : CH . CH_3$, and from this three alcohols might be derived:

1. $CH(OH) : CH . CH_3$, a secondary alcohol,
2. $CH_2 : C(OH) . CH_3$, a tertiary ,,
3. $CH_2 : CH . CH_2(OH)$, a primary ,,

Of these, the *third* formula represents allyl alcohol, which in many respects behaves like any other primary alcohol, but differs from methyl alcohol and its homologues in being unsaturated (see p. 28). On the one hand, as a primary alcohol, it yields an aldehyde and then an acid when oxidised, while as an unsaturated compound it is able to combine directly with chlorine or bromine.

Allyl Alcohol, $C_3H_5.OH$, is obtained by distilling a mixture of glycerine, $C_3H_5(OH)_3$, with formic acid (oxalic acid may be substituted for this, but as it decomposes under these conditions into formic acid and CO_2, the reaction is practically the same). The formic acid is oxidised to CO_2 and water:

$$C_3H_5(OH)_3 + HCO_2H = C_3H_5.OH + CO_2 + 2H_2O.$$
Glycerine.   Formic acid.   Allyl alcohol.

The following is the usual method for preparing allyl alcohol :—

Four parts of glycerine and one of crystallised oxalic acid are placed in a retort and gradually heated. At first much CO_2 is evolved, and dilute formic acid distils over. When the temperature of the mixture reaches 190° the receiver is changed, and impure allyl alcohol is obtained as the distillate. This is purified by fractional distillation, and freed from water by treatment with anhydrous baryta. Pure allyl alcohol boils at 96°.

Allyl alcohol is a colourless liquid which, like all the allyl compounds, has an irritating, unpleasant smell. As an unsaturated body it combines directly with Cl_2 or Br_2 to form derivatives of propyl alcohol :

$$C_3H_5.OH + Br_2 = C_3H_5Br.OH.$$
Allyl alcohol.   Dibromo-propyl alcohol.

As a primary alcohol allyl alcohol yields, when carefully oxidised, first an aldehyde—allyl aldehyde or acrolein—and then an acid—acrylic acid :

$$CH_2:CH.CH_2OH + O = CH_2:CH.CHO + H_2O$$
Allyl alcohol.   Acrolein.

$$CH_2:CH.CHO + O = CH_2:CH.CO_2H.$$
Acrolein.   Acrylic acid.

Acrolein, $C_2H_3.CHO$, is also produced when glycerine or fats (which are compounds of glycerine) are heated to decomposition. It is best obtained by distilling glycerine to which twice its weight of $KHSO_4$ has been added:

$$C_3H_5(OH)_3 = C_2H_3.CHO + 2H_2O.$$
$$\text{Glycerine.} \qquad \text{Acrolein.}$$

Acrolein is a volatile liquid with an extremely irritating odour. Its chemical behaviour is fairly summed up in the statement that it is an unsaturated aldehyde.

Acrylic Acid, $C_2H_3.CO_2H$, is best obtained from acrolein by boiling it with water and oxide of silver:

$$C_2H_3.CHO + Ag_2O = C_2H_3.CO_2H + 2Ag.$$
$$\text{Acrolein.} \qquad \text{Acrylic acid.}$$

Acrylic is a well-marked acid. It is of course an unsaturated body and, as such, combines readily with chlorine, bromine, etc., to form derivatives of propionic acid:

$$C_2H_3.CO_2H + Br_2 = C_2H_3Br_2.CO_2H.$$
$$\text{Acrylic acid.} \qquad \text{Dibrom-propionic acid.}$$

It is a liquid similar to acetic acid in appearance and smell.

Allyl alcohol forms ethereal salts with acids, but of these the following are alone of sufficient importance to be mentioned here :—

Allyl Iodide, C_3H_5I, is a colourless liquid, which can be obtained from the alcohol by the action of HI:

$$C_3H_5.OH + HI = C_3H_5I + H_2O,$$
$$\text{Allyl alcohol.} \qquad \text{Allyl iodide.}$$

or more conveniently from glycerine by the action of phosphorus and iodine, a reaction which may be supposed to occur in the two following stages:

$$C_3H_5(OH)_3 + PI_3 = C_3H_5I_3 + H_3PO_3$$
$$\text{Glycerine.} \qquad \text{Glyceryl iodide.}$$

$$C_3H_5I_3 = C_3H_5I + I_2$$
$$\text{Glyceryl iodide.} \quad \text{Allyl iodide.}$$

The experimental details of the second method of preparation are as follows:

A quantity of glycerine is freed from water by heating in an open dish for at least half an hour to such a temperature that the liquid is near its boiling point and evolves abundant fumes. The anhydrous glycerine must be placed in a well-stoppered bottle while still warm.

A tubulated retort is fitted with a cork and connected with an apparatus for generating CO_2, so that a slow current of that gas can be passed through the retort during the whole experiment; 150 grams of the anhydrous glycerine is then placed in the retort, along with 100 grams of powdered iodine; 60 grams of yellow phosphorus is weighed out and cut into small pieces, which are taken up one by one at the end of a knife, dried between filter-paper, and introduced through the tubulus into the retort. A violent reaction occurs as each piece of phosphorus is added, and impure allyl iodide distils over; it is washed with a solution of soda, separated by means of a tap-funnel from the soda, dried by contact with a few pieces of fused calcium chloride, and re-distilled. Pure allyl iodide boils at 101° C.

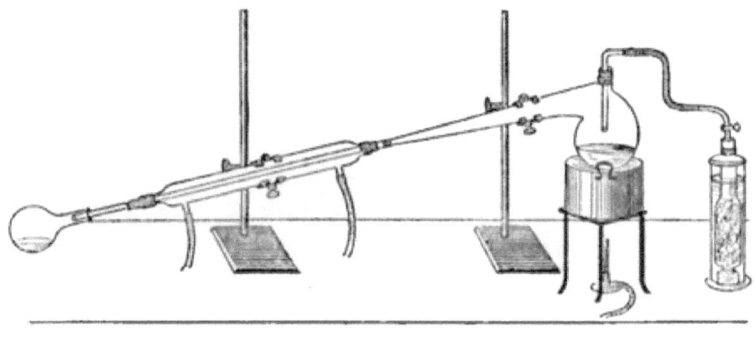

Fig. 31.—Preparation of Allyl Iodide.

Allyl Sulphide, $(C_3H_5)_2S$, is the chief constituent of *oil of garlic*, which is obtained by distilling garlic with steam, and gives that plant its characteristic smell and taste. It can be prepared artificially by the action of allyl iodide upon potassium sulphide:

$$K_2S + 2C_3H_5I = 2KI + (C_3H_5)_2S.$$
$$\text{Allyl iodide.} \qquad\qquad \text{Allyl sulphide.}$$

Allyl Iso-thiocyanate, $C_3H_5 . NCS$, is present in *oil of* *mustard*, obtained by distillation of mustard seeds. It can be prepared artificially by the action of allyl iodide upon potassium thiocyanate KCNS:

$$KCNS + C_3H_5I = KI + C_3H_5 . NCS.$$
Oil of mustard.

It is a liquid with the strong penetrating odour and taste of the natural "oil of mustard."

Questions on Chapter XVI

1. How can allyl alcohol be obtained from glycerine? What reactions stamp allyl alcohol as an unsaturated compound?

2. By what reactions is it possible to prepare acrylic acid from glycerine?

3. What reasons have we for regarding allyl alcohol as an unsaturated primary alcohol?

4. Give the formulæ and systematic names of (*a*) oil of mustard, (*b*) oil of garlic. How can each be prepared artificially?

CHAPTER XVII

GLYCERINE AND ITS COMPOUNDS

Glycerine is contained in fats and fatty oils combined with organic acids in the form of ethereal salts. When these compounds are heated with alkalies in the preparation of soap the glycerine is set free, and when the soap is separated by addition of salt from the liquor in which the glycerine is contained, this latter can be easily recovered. In many soaps now manufactured the water and glycerine are not separated from the true soap, but the whole is allowed to cool, when it solidifies to a mass naturally less firm than a pure soap and less durable, but pleasanter to use and far more profitable to manufacture. Soap manufacture is accordingly not a very important source of glycerine; far more is obtained in the preparation of stearic acid for candles. The best process conducts the saponification of the fat by means of superheated steam with the use of a small proportion of lime. Stearic acid (mixed with other fatty acids) and glycerine are produced:

$$\text{Fat} + \text{Water} = \text{Stearic Acid} + \text{Glycerine}.$$

Glycerine is found by analysis to have the formula $C_3H_8O_3$. It behaves as a *trihydric alcohol*, and yields ethereal salts with various acids, in which three acid groups are introduced into the glycerine molecule; this leads us to write the formula as $C_3H_5(OH)_3$.

Glycerine when perfectly pure forms colourless crystals which melt at 17° C., about the ordinary temperature of a room. It is, however, very hygroscopic, and a trace of water is sufficient to convert it into a syrupy liquid; this has a sweet taste, and is sometimes added to wine to give it body and sweetness. It is also used as a cosmetic and for keeping leather articles soft and pliable; it is the starting-point in the manufacture of nitro-glycerine and dynamite.

When distilled under the ordinary pressure, glycerine is largely decomposed, acrolein being one of the principal products:

$$C_3H_5(OH)_3 = C_3H_4O + 2H_2O,$$
$$\text{Glycerine.} \qquad \text{Acrolein.}$$

but under diminished pressure or in a current of superheated steam it can be distilled without decomposition.

Glycerine can be prepared synthetically from allyl tribromide $CH_2Br.CHBr.CH_2Br$, just as glycol from $CH_2Br.CH_2Br$ and ethyl alcohol from C_2H_5Br; its constitutional formula is $CH_2(OH).CH(OH).CH_2(OH)$, and when oxidised it yields first glyceric and then tartronic acids:

$$\begin{array}{ccccc}
CH_2.OH & & CO_2H & & CO_2H \\
| & & | & & | \\
CH.OH & \longrightarrow & CH.OH & \longrightarrow & CH.OH \\
| & & | & & | \\
CH_2.OH & & CH_2.OH & & CO_2H \\
& & \text{Glyceric acid.} & & \text{Tartronic acid.}
\end{array}$$

The most important compound of glycerine is the nitrate, generally known as *nitro-glycerine*; this is obtained by the action upon glycerine of a mixture of concentrated sulphuric and nitric acids; the product is added to water when the nitro-glycerine separates as an oil, which has to be thoroughly washed before being stored or worked up into dynamite, as otherwise the traces of acid left in the oil render it liable to explode on very slight provocation.

Nitro-glycerine has the constitution $C_3H_5(NO_3)_3$; it is the nitrate of the tri-valent radicle C_3H_5 (glyceryl), and its formation is represented by the equation:

$$C_3H_5(OH)_3 + 3HNO_3 = C_3H_5(NO_3)_3 + 3H_2O.$$
$$\text{Glycerine.} \qquad\qquad \text{Glyceryl nitrate}$$
$$\text{or nitro-glycerine.}$$

The sulphuric acid used in its manufacture merely aids the

action of the nitric acid by combining with the water produced.

By boiling with water and an alkali, nitro-glycerine (like other ethereal salts) is converted into the alcohol and acid from which it was formed:

$$C_3H_5(NO_3)_3 + 3KOH = C_3H_5(OH)_3 + 3KNO_3.$$
Nitro-glycerine. Glycerine.

Nitro-glycerine, like most other similar compounds (see gun-cotton, p. 126), decomposes very readily when heated or exposed to sudden shock. The substance contains more oxygen than is required to burn up the carbon and hydrogen contained in it:

$$2C_3H_5(NO_3)_3 = 6CO_2 + 5H_2O + 3N_2 + O\ ;$$

hence no oxygen from outside is required, and nitro-glycerine can burn or explode when cut off from contact with air. Moreover, the oxygen with which the carbon and hydrogen combine is present in the same molecule with them, and in consequence the change represented in the above equation takes place with extreme rapidity and suddenness when once started. The heat produced in the reaction is therefore also very suddenly developed, and the destructive power of nitro-glycerine is far in excess of that of a quantity of gunpowder, which in burning would give out the same total amount of heat.

Nitro-glycerine is a very dangerous substance to handle, as even when very carefully prepared it requires only a slight shock to make it explode. This disadvantage is largely removed in *dynamite*, which is a mixture of nitro-glycerine with very fine siliceous earth. More recently this has been almost superseded by *blasting-gelatine*, a jelly-like solid obtained by dissolving gun-cotton in nitro-glycerine, which is even safer to handle, and can, by varying the proportions, be made in different grades of violence according to the purpose intended. By addition of camphor, or other appropriate substance to this mixture, a material is obtained of sufficiently moderate explosive power to be used in ordinary firearms—the modern smokeless powder.

Of some theoretical interest are the **chlorhydrins**, compounds obtained from glycerine by the action of HCl or of PCl_5; in these the hydroxyl groups of the $C_3H_5(OH)_3$ are more or less completely replaced by chlorine; they are ethereal salts of the trihydric alcohol glycerine and hydrochloric acid.

There are two **mono-chlorhydrins**, (a) $CH_2(OH).CH(OH).CH_2Cl$ and (β) $CH_2(OH).CHCl.CH_2(OH)$, of which the first is obtained by the action of HCl on glycerine.

Of the two **di-chlorhydrins** one has the formula $CH_2Cl.CH(OH).CH_2Cl$, and is obtained by the action of HCl on glycerine; the other one is $CH_2Cl.CHCl.CH_2(OH)$, and is the addition product of allyl alcohol (see p. 111) and Cl_2.

Trichlorhydrin, $C_3H_5Cl_3(CH_2Cl.CHCl.CH_2Cl)$, is the final result of the action of HCl (or better, PCl_5) upon glycerine; it is one of the five possible isomeric trichloropropanes, and is a liquid with a smell like chloroform.

QUESTIONS ON CHAPTER XVII

1. What is the chemical constitution of fat? How are the fats worked up in the manufacture of glycerine?

2. What happens to glycerine (a) when heated in the air, (b) when treated with a mixture of nitric and sulphuric acids?

3. What chemical changes occur when nitro-glycerine (a) explodes, (b) is heated gently with dilute caustic soda?

4. How is the dangerous violence of nitro-glycerine modified in several modern explosives?

CHAPTER XVIII

THE CARBOHYDRATES

The **Carbohydrates** are a class of bodies of extreme importance, especially in plant life; not only are they the chief constituents of all plants, but they are also present in many animal tissues.

All the carbohydrates are composed of the three elements, carbon, hydrogen, and oxygen, and of these elements the two latter are present in the proportion in which they combine to form water; their number is very large, and their accurate investigation is surrounded with such difficulties that only in recent years has much real knowledge of their chemistry been gained.

The chief difficulty was that no reagent was known with which the carbohydrates would yield well-characterised products; the compounds which they, as aldehyde and ketone-alcohols, form with phenyl-hydrazine can, however, for the most part be distinctly and easily recognised, and it is by their help that much of our recent knowledge in this field has been won.

THE GLUCOSES

The first family of the carbohydrates to be considered is the **Glucoses**; these have the empirical formula CH_2O, and most of them have the molecular formula $C_6H_{12}O_6$, as has been proved by the application of Raoult's method for the determination of molecular weights (see p. 16).

There are, however, bodies known of the molecular formulæ $C_5H_{10}O_5$ (arabinose) and $C_7H_{14}O_7$ (heptose), which are best included in this group.

The glucoses have a sweet taste, though less sweet than cane-sugar; they are easily soluble in water, and at once reduce Fehling's solution (p. 62); they also readily ferment under the influence of yeast. The various glucoses differ from one another in crystalline form, in their solubility in various reagents, and in other properties; their isomerism cannot be satisfactorily accounted for by the ordinary theories of the structure of carbon compounds, and its fuller explanation is undoubtedly to be found in the application of Van't Hoff's theory of the tetrahedral carbon atom (see p. 108). In view of this, special importance is attached to the varying power of the different glucoses to rotate the plane of polarisation of light.

Chemically the glucoses are, in the first place, *alcohols;* they (at least those of the formula $C_6H_{12}O_6$) contain five OH groups (each of which can be replaced by acetyl upon treatment with acetic anhydride, see p. 79). In the second place, the reducing power of the glucoses leads to the conclusion that they are *aldehydes* or *ketones* as well as alcohols; they contain therefore five OH groups and one CHO or CO group.

This CHO or CO group can be converted by reduction into a sixth alcohol group—CH_2OH or CHOH. We thus obtain by reduction of a glucose a hexhydric alcohol, which may be regarded as the parent of that particular glucose. The alcohol obtained has in each case the formula $C_6H_8(OH)_6$, but while two of the glucoses (dextrose and levulose) yield *mannitol*, a third (galactose) yields an isomer of that substance, viz. *dulcitol.*

Mannitol is also contained in manna, and is present in many plants, as is also the isomeric *dulcitol*. Both are derivatives of normal hexane, C_6H_{14}, and their isomerism is to be explained by Van't Hoff's theory of the tetrahedral carbon atom.

Dextrose, $C_6H_{12}O_6$, is present in many fruits, and also in honey. It rotates the plane of polarisation to the right.

Dextrose is formed in the hydrolysis (splitting up of compounds by addition of water) of many other carbohydrates. The hydrolysis is effected by heating with water under pressure, or more easily by boiling with a dilute mineral acid. Thus we have the following reactions :—

$$\text{Cane-sugar} + H_2O = \text{Dextrose} + \text{Levulose}$$
$$\text{Starch} + H_2O = \text{Dextrose}$$

The dextrose of commerce is prepared by treating starch with boiling dilute sulphuric acid under pressure. The solution is freed from sulphuric acid by adding calcium carbonate and filtering from the calcium sulphate; it is then evaporated, and leaves a tough non-crystalline mass.

Levulose, $C_6H_{12}O_6$, occurs along with dextrose in fruits and honey, and the "invert-sugar" obtained by the action of dilute acids on cane-sugar is a mixture of equal parts of dextrose and levulose.

Levulose rotates the plane of polarisation more strongly than dextrose, but to the left.

The dextrose and levulose which are present together in honey and in invert-sugar can be partially separated by washing the mixture with cold alcohol. The more soluble levulose is thus removed dissolved in the alcohol, and the less soluble dextrose remains for the most part undissolved.

Galactose, $C_6H_{12}O_6$, is formed along with dextrose in the hydrolysis of milk-sugar :

$$\text{Milk-sugar} + H_2O = \text{Dextrose} + \text{Galactose}.$$

Unlike dextrose and levulose, galactose does not ferment with yeast. When reduced it yields dulcitol :

	Rotation of polarised light.	Action of yeast.	Reduction product.
Dextrose	To right	Ferments	Mannitol
Levulose	To left	Ferments less rapidly than glucose	Mannitol
Galactose	To right	Does not ferment	Dulcitol

CANE-SUGAR GROUP OR BIOSES

The members of this group are made up of two molecules of glucose united together with elimination of a molecule of water. When hydrolysed (see above) they take up water to form two molecules of glucose. The formula is $C_{12}H_{22}O_{11}$, and the following table indicates the relation of the most important members of the group to the glucoses:

$$\text{Cane-sugar} + H_2O = \text{Dextrose} + \text{Levulose}$$
$$\text{Milk-sugar} + H_2O = \text{Dextrose} + \text{Galactose}$$
$$\text{Maltose} \quad + H_2O = \text{Dextrose} + \text{Dextrose}$$

The Bioses are not so strong reducing agents as the Glucoses. None of them is able to reduce Fehling's solution in the cold,

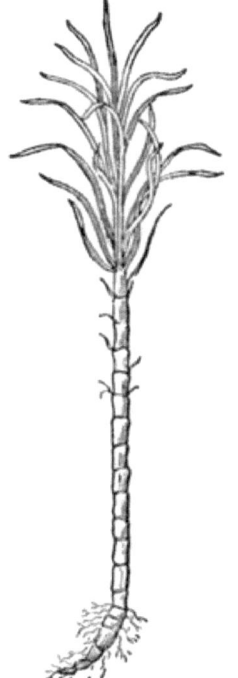

Fig. 32.—Sugar-cane.
Yield of canes per acre, 30-40 tons, containing about 5 tons of sugar.

Fig. 33.—Sugar-beet.
Yield of beet per acre, 15-20 tons, containing about 2 tons of sugar.

but maltose does so readily when heat is applied. The others reduce it only very slowly even when boiled.

Cane-sugar, $C_{12}H_{22}O_{11}$, is present in the sap of many plants, especially the sugar-cane and the beet-root. In order to obtain the sugar the sap is extracted either by crushing and pressure, or by cutting into thin slices and soaking in water. The juice is purified by filtration and other processes, and is then evaporated in vacuum-pans until sugar separates out from the juice on cooling.

A portion only of the sugar is thus obtained in the crystalline state, the remainder is left in the form of a thick syrup after the crystals have been removed, and the sugar in it is prevented by impurities from crystallising. These "molasses" may either be fermented and converted into spirit (rum), or by certain modern processes the impurities may be separated and the sugar obtained in the solid form.

In one of these, the diffusion process, the syrup is put into what are practically huge bags made of parchment paper, and these bags are placed in pure water. The sugar of the molasses diffuses through the pores of the parchment paper faster than the impurities which are mixed with it, and there is thus obtained in the liquor surrounding the bags a solution of sugar sufficiently pure to yield a crystalline product on evaporation.

The other process depends on the formation of a nearly insoluble compound with lime, having the composition $C_{12}H_{22}O_{11} \cdot 3CaO$. This is precipitated from the molasses by addition of powdered quick-lime, and after being purified by washing, is decomposed by passing a stream of CO_2 through water with which the "lime saccharate" is mixed. The lime is separated as calcium carbonate, and a weak syrup of pure sugar is obtained, which can readily be concentrated by evaporation.

Cane-sugar crystallises in large monoclinic prisms (sugar-candy). It is very soluble in water, but is easily crystallised from its solutions by evaporation unless the presence of impurities interferes. Solutions rotate the plane of polarisation of light to the right. On boiling with a dilute acid cane-sugar is converted into a mixture of dextrose and levulose :

$$\underset{\text{Cane-sugar.}}{C_{12}H_{22}O_{11}} + H_2O = \underset{\text{Dextrose.}}{C_6H_{12}O_6} + \underset{\text{Levulose.}}{C_6H_{12}O_6} ;$$

and as levulose has a higher rotatory power than dextrose, the mixture of the two thus obtained rotates to the left; this inversion is the origin of the name invert-sugar, which is applied to the product thus obtained.

When heated, cane-sugar melts at 160° C., and if then allowed to cool solidifies to a semi-transparent mass ("barley-sugar"), which is devoid of crystalline structure. On long standing this gradually becomes crystalline again. If heated to about 200° C., cane-sugar is changed into a brown substance known as "caramel," or "burnt-sugar," which is used as colouring matter by cooks.

Cane-sugar when subjected to the influence of the growing yeast-plant is first changed into a mixture of dextrose and levulose. As soon as any considerable quantity of these glucoses has been formed alcoholic fermentation sets in, following chiefly the equation

$$\underset{\text{Glucose.}}{C_6H_{12}O_6} = \underset{\text{Ethyl alcohol.}}{2C_2H_6O} + 2CO_2.$$

Milk-sugar, $C_{12}H_{22}O_{11}$, is present in milk, and remains dissolved in the whey after the casein has been separated in the manufacture of cheese. It is less soluble in water than cane-sugar, and much less sweet. Its solutions rotate the plane of polarisation to the right.

Milk-sugar does not easily ferment with yeast, but by the action of certain bacteria it readily ferments with production of lactic acid:

$$\underset{\text{Milk-sugar.}}{C_{12}H_{22}O_{11}} + H_2O = \underset{\text{Lactic acid.}}{4C_3H_6O_3}.$$

This is the change which occurs when milk turns sour.

As has been already mentioned, the hydrolysis of milk-sugar yields dextrose and galactose:

$$\underset{\text{Milk-sugar.}}{C_{12}H_{22}O_{11}} + H_2O = \underset{\text{Dextrose.}}{C_6H_{12}O_6} + \underset{\text{Galactose.}}{C_6H_{12}O_6}.$$

Maltose, $C_{12}H_{22}O_{11}$, is contained in malt, having been produced by the action of a certain ferment—diastase—upon the starch present in the barley or other grain which has been malted.

Upon hydrolysis—boiling with a dilute acid—maltose yields dextrose only :

$$C_{12}H_{22}O_{11} + H_2O = 2C_6H_{12}O_6.$$
$$\text{Maltose.} \qquad\qquad \text{Dextrose.}$$

It resembles the glucoses much more closely than do cane- and milk-sugar; thus it ferments quickly (*i.e.* without previous conversion into glucoses) with yeast, and reduces Fehling's solution readily when warmed with it.

THE CELLULOSE GROUP

In this group we include a number of carbohydrates whose constitution is less understood even than that of the glucoses and bioses. It is very probable that their molecular weights are very high, but it has not yet been found possible to determine their real values, and we can only give the empirical formulæ. The most important members of the group are *starch*, *dextrin*, and *cellulose*—all $C_6H_{10}O_5$—and the *gums*, whose probable formula is $C_5H_{10}O_5$.

Starch, $C_6H_{10}O_5$, is the form in which very many plants store up their reserves of food. It is largely present in many roots and seeds, as the following table will show :—

	Per cent of starch.
Potatoes	20
Wheat, maize	60
Rice	70

Starch is also a very important food for animals; and arrow-root, sago, and tapioca are nearly pure starch extracted from certain plants. The separation of starch from the other constituents of the plants is effected by beating them with water into a thin pulp, which is then filtered through fine sieves. The fibrous matter is kept back, and the milky liquid which runs through deposits the starch on standing. This is then collected and dried.

Starch is really insoluble in water, but when boiled with it yields a liquid which can be filtered without separating the starch. This is, however, merely present in a very fine state of subdivision, forming what is called a "colloidal" solution. The starch in it is unable to pass through a membrane of parchment paper, whereas substances in real solution are able slowly to diffuse through such a membrane. Neither has the starch any effect on the freezing-point of the water containing it (see p. 16).

When heated to about 200° C., starch is changed into dextrin.

Very characteristic of starch is the intensely blue compound which it forms with iodine. This furnishes a very sensitive test either for starch or for free iodine. The blue colour disappears when sufficient heat is applied, but reappears on cooling.

Dextrin, $C_6H_{10}O_5$, is obtained by simply heating starch to about 200° C., or by boiling it with dilute acids. Dextrin is used as a substitute for gum. It is not coloured blue by iodine.

Cellulose, $C_6H_{10}O_5$, is the chief constituent of the cell-walls of plants; wood is chiefly cellulose, while cotton-wool and filter-paper are nearly pure cellulose. This is insoluble in all ordinary solvents, but concentrated sulphuric acid dissolves it, and the solution when diluted and boiled yields first dextrin and then dextrose.

The exact chemical constitution of cellulose is matter for future investigation. It appears, however, to contain three-fifths of its oxygen in the form of hydroxyl groups OH, as we find that by the action of acids ethereal salts of cellulose may be prepared in which three acid groups are introduced into the formula $C_6H_{10}O_5$; *the real molecular formula of cellulose is unknown*, but it is more convenient to regard these ethereal salts as derived from the doubled formula $C_{12}H_{20}O_{10}$, in which, of course, there are six hydroxyls.

The most important of these salts are the nitrates; these are prepared by treating cellulose (cotton-wool) with strong nitric acid, the action being aided by the addition of concentrated sulphuric acid. When the strongest acids are employed the product obtained is *gun-cotton*, which is found to be cellulose hexa-nitrate:

$$C_{12}H_{14}O_4(OH)_6 + 6HNO_3 = C_{12}H_{14}O_4(NO_3)_6 + 6H_2O.$$
Cellulose. Gun-cotton.

This material is a violent explosive, and is prepared by steeping cotton-wool for a few minutes in a cold mixture of the strongest nitric acid with two or three times its weight of concentrated sulphuric acid. When thoroughly freed from acid by washing, gun-cotton is comparatively quite safe to handle, and may even be set fire to without anything more violent than a rather quick combustion taking place ; but when subjected to the shock set up by exploding a small charge of fulminate embedded in the gun-cotton, the molecules of the latter break down suddenly, and a powerful explosion results ; the rearrangement of atoms which then occurs may be roughly represented by the following equation :

$$C_{12}H_{14}O_4(NO_3)_6 = 3N_2 + 7H_2O + 9CO + 3CO_2.$$
Gun-cotton.

Pyroxylin is a less highly nitrated cellulose, chiefly the tetra-nitrate ; it is prepared with a somewhat weaker nitric acid. Its solution in a mixture of alcohol and ether is the *collodion* which is largely used in photography (wet-plate process), and in surgery for covering wounds with a thin flexible film which prevents access of air.

QUESTIONS ON CHAPTER XVIII

1. What are the chief members of the group of "glucoses"; what is their formula, and what explanation of their isomerism may be advanced?
2. What is the action upon dextrose of (*a*) Fehling's solution, (*b*) yeast, (*c*) acetic anhydride?
3. What two substances are present in largest quantity in honey? How do they differ from one another?
4. What products are obtained by the action of boiling dilute acids upon (*a*) cane-sugar, (*b*) milk-sugar, (*c*) maltose?
5. Describe the preparation of gun-cotton, and give its chemical constitution.

CHAPTER XIX

UREA AND URIC ACID

Urea, $CO(NH_2)_2$, is one of the most important waste-products of the animal economy; the food which animals consume is converted during its passage through the blood and tissues of the body chiefly into urea, carbon dioxide, and water. The urea is secreted along with a considerable proportion of the water by the kidneys, and it was from urine that this substance was first obtained in 1773.

Urea thus obtained and afterwards carefully purified was found by analysis to have the composition CON_2H_4. This is also the composition of ammonium cyanate, $(NH_4)NCO$, and though that body is itself quite distinct from, and isomeric with, urea, in 1828 Wöhler made the very important discovery that a solution of ammonium cyanate in water yields urea on evaporation. The great readiness with which this change occurs while indicating that urea is the more stable of the two isomers also seems to show that they are not very different in constitution. Several synthetical methods which have since been discovered for the preparation of urea show that it may be regarded as the amide of carbonic acid, $CO{<}{NH_2 \atop NH_2}$. Thus just as the amide of acetic acid (acetamide) can be got by the action of ammonia on acetyl chloride:

$$CH_3 . COCl + NH_3 = CH_3 . CONH_2 + HCl,$$
Acetyl chloride.   Acetamide.

so urea can be obtained by the action of ammonia on carbonyl chloride $COCl_2$:

$$COCl_2 + 2NH_3 = CO(NH_2)_2 + 2HCl.$$
Carbonyl chloride. Carbamide or urea.

The most convenient way of preparing urea is by evaporating a solution in water of potassium cyanate and ammonium sulphate mixed in equivalent proportions; the potassium cyanate is easily obtained by heating potassium ferrocyanide with manganese dioxide.

EXPT. 19. Heat four parts potassium ferrocyanide with two parts of MnO_2 in a clay crucible, extract the cooled melt with water, add three parts of ammonium sulphate, and evaporate to dryness. Potassium sulphate and urea are left, and may be separated by extraction with alcohol, in which the urea only is soluble.

Urea crystallises in rhombic prisms which are easily soluble in water. It is a mon-acid base, and forms salts of which the nitrate $CON_2H_4 . HNO_3$ is very sparingly soluble in water containing nitric acid, and may therefore be used as a means of detecting urea in solutions not too dilute.

Like other amides urea is decomposed on boiling with dilute alkalies, and ammonia is given off:

$$CO(NH_2)_2 + H_2O = CO_2 + 2NH_3.$$
Urea.

Another important reaction of urea is its behaviour when treated with bromine and caustic soda (sodium hypobromite); it is then oxidised to CO_2 and water while the nitrogen is given off as such:

$$CON_2H_4 + 3NaOBr = CO_2 + 2H_2O + N_2 + 3NaBr.$$

EXPT. 20. Put some solution of urea in a boiling tube, add caustic soda and bromine water; notice that a gas is given off in bubbles, and by testing with a match show that it puts out the flame. (The gas cannot be CO_2, because the solution contains excess of alkali.) The experiment can be so arranged that the nitrogen may be collected and measured; from its amount that of the urea can be calculated, and on this a method for estimating urea is based. It must, however, be remembered that many other nitrogen compounds also give off their nitrogen when treated with a hypobromite.

Uric Acid, $C_5H_4N_4O_3$ (5443), may be regarded as a less completely oxidised result of the digestive and absorptive processes than urea. Uric acid is present only in small quantity in the urine of man, but in certain abnormal conditions of the body it is more largely produced, usually with very unpleasant consequences. Both uric acid and its salts are soluble only with difficulty in water, hence they are difficult to remove when produced in the body in any considerable quantity, and either gout, in which accumulations of urates occur in various parts of the body, or other disturbances of the healthy procedure occur.

In some animals, on the other hand, especially birds and reptiles, uric acid is largely secreted, and both guano (which is produced by sea-birds) and the excreta of serpents contain considerable quantities, and from either of these sources the acid may readily be prepared.

If guano is used it is best boiled with a solution of borax (1 to 100 of water), in which uric acid is fairly soluble. Addition of hydrochloric acid to the filtered solution precipitates the bulk of the uric acid present.

Uric acid is a white powder, soluble only very slightly in pure water, but more readily in water containing certain salts in solution. It is a weak di-basic acid, but the best characterised salts are those with only one equivalent of metal, such as $C_5H_3KN_4O_3$, potassium urate; they are all very slightly soluble in water.

To test a substance for the presence of uric acid a few drops of dilute nitric acid are added to it, and then evaporated on the water bath; if a yellow residue is left which is coloured purple by addition of ammonia, we may conclude that uric acid was contained in the substance examined.

Questions on Chapter XIX

1. How can urea be prepared from potassium ferrocyanide?

2. What is the action upon urea of (*a*) sodium hypobromite, (*b*) boiling caustic soda solution?

3. What products are formed by the action of ammonia upon (*a*) carbonyl chloride, (*b*) acetyl chloride?

4. From what sources can uric acid be obtained? Write down the formulæ of uric acid and of potassium urate.

CHAPTER XX

THE CYANOGEN COMPOUNDS

THE cyanogen compounds include a large number of substances which are alike in containing the monovalent radicle *cyanogen* – CN, made up, as its formula shows, of one atom each of tetravalent carbon and trivalent nitrogen: $-C\equiv N$. Sometimes the special symbol Cy is used to denote the cyanogen radicle.

The starting-point in the preparation of the various cyanogen compounds is potassium ferrocyanide, or "yellow prussiate of potash," but as the composition of this substance is somewhat complex it is better to begin with other and simpler bodies.

Cyanogen, C_2N_2, a compound whose molecule is formed of two cyanogen radicles united together (just as free chlorine or hydrogen is Cl_2 or H_2), is made by heating mercuric cyanide to a red heat, when it decomposes into mercury and cyanogen:

$$Hg(CN)_2 = Hg + C_2N_2.$$

It is a poisonous gas with a characteristic smell, and burns in air with a peculiar ("peach-blossom colour") flame; its mixture with oxygen explodes violently on application of a flame:

$$C_2N_2 + 2O_2 = 2CO_2 + N_2.$$

Cyanogen is readily soluble in water, and must therefore be collected over mercury.

Chemically cyanogen behaves as the "nitrile" of oxalic acid. It can be obtained from the amide of oxalic acid—oxamide $(CONH_2)_2$—by withdrawing water (action of P_2O_5):

$$\begin{array}{l} \text{CONH}_2 \\ | \quad\quad\quad -2\text{H}_2\text{O} = \text{C}_2\text{N}_2\,; \\ \text{CONH}_2 \end{array}$$

and the inverse reaction can be brought about by allowing a solution of cyanogen in water or dilute acid to stand for several days:

$$\text{C}_2\text{N}_2 + 2\text{H}_2\text{O} = \text{C}_2\text{O}_2(\text{NH}_2)_2.$$

Compare the relation of methyl cyanide (acetonitrile) CH_3CN to acetamide, pp. 89 and 134.

Hydrocyanic Acid, HCN, or "Prussic Acid," is now most largely prepared by the action of boiling dilute sulphuric acid upon potassium ferrocyanide:

$$2\text{K}_4\text{FeCy}_6 + 3\text{H}_2\text{SO}_4 = 3\text{K}_2\text{SO}_4 + \text{K}_2\text{Fe}_2\text{Cy}_6 + 6\text{HCN}.$$

By this method a solution of hydrocyanic acid in water is obtained, from which the anhydrous acid can be prepared by passing the vapours through tubes containing calcium chloride or other suitable dehydrating agent.

An older method of preparing the dilute acid is from *amygdalin*, a compound present in bitter almonds, laurel leaves, and parts of various other plants. The amygdalin, when the leaves, etc., steeped in water, are exposed to the air, usually undergoes a fermentation which results in the formation of hydrocyanic acid, oil of bitter almonds (benzaldehyde, see Part II.), and sugar. The hydrocyanic acid is then easily obtained by distillation.

The salts of hydrocyanic acid—the cyanides—are formed whenever carbon and nitrogen come in contact with a strong base at a high temperature. The nitrogen may be supplied in the free state, or may be present in combination with other elements. Thus potassium cyanide is formed when nitrogen is passed over a heated mixture of potash and powdered coal, and cyanides are always formed in distilling coal for the manufacture of coal-gas from the joint interaction of ammonia with the nitrogen and carbon present in the coal. The two chief sources of the cyanides (which are largely manufactured for use in electro-plating, for making Prussian blue, and other purposes) are to be associated with this method of formation. They are—

1. Potassium ferrocyanide, yellow prussiate of potash, which is made by carbonising nitrogenous animal refuse (horn, leather scraps, etc.) and heating the residue, which, though chiefly carbon, still contains a considerable proportion of nitrogen, with caustic potash and iron filings.

2. An important source of cyanides is now found in the by-products of the manufacture of coal-gas. The cyanides formed during the destructive distillation of the coal are retained chiefly in the lime-purifiers, and are extracted from the spent lime by treatment with quicklime at steam heat. This decomposes the insoluble cyanogen compounds present in the spent lime, and converts them into soluble calcium ferrocyanide.

Cyanides are now also recovered from the by-products of other manufactures—blast-furnaces, coke-ovens, etc.

Other reactions, of theoretical interest only, by which hydrocyanic acid or its salts can be obtained, are—

1. The action of the electric discharge upon a mixture of acetylene and nitrogen:

$$C_2H_2 + N_2 = 2HCN.$$

2. The action of ammonia upon chloroform (in the presence of caustic potash)

$$NH_3 + CHCl_3 = 3HCl + HCN.$$

Pure hydrocyanic acid, free from water, is a colourless volatile liquid, with a strong smell and intensely poisonous properties. It is a well-marked acid, but its salts with the alkaline metals are easily decomposed, even carbon dioxide being sufficiently powerful to liberate the acid from potassium or ammonium cyanides; hence it is that these substances always smell of hydrocyanic acid when exposed to the air. The cyanides of the heavy metals are, on the other hand, much more stable, silver cyanide being unattacked even by the strong acids.

The solution of hydrocyanic acid in water readily decomposes with formation of ammonium formate and other substances. A similar change occurs more readily by the action of dilute mineral acids, showing that hydrocyanic acid may be regarded as the nitrile of formic acid:

$$\underset{\text{Hydrogen cyanide.}}{HCN} + 2H_2O = \underset{\text{Formic acid.}}{HCO_2H} + NH_3,$$

with which compare

$$CH_3CN + 2H_2O = CH_3 \cdot CO_2H + NH_3.$$
Methyl cyanide or "acetonitrile."   Acetic acid.

Potassium Cyanide, KCN, is manufactured by heating potassium ferrocyanide in iron vessels until decomposition occurs according to the equation:

$$K_4FeC_6N_6 = 4KCN + FeC_2 + N_2.$$

The potassium cyanide is separated from the iron-carbide by extracting the mass with water; the solution is evaporated, and the residue, after being fused, is brought into market in lumps or sticks.

Perfectly pure potassium cyanide is best obtained by passing vapours of HCN into a solution of KOH in alcohol.

Potassium cyanide is very soluble in water, and is extremely poisonous. It is largely used in electro-plating for preparing the solutions of gold or silver, and in the gold-fields for dissolving the gold from the quartz containing it. It is used in the laboratory as a reducing agent in blow-pipe work.

Mercuric Cyanide, $Hg(CN)_2$, is prepared by boiling Prussian blue with water and mercuric oxide, or by dissolving mercuric oxide in hydrocyanic acid. It is fairly soluble in water, is very poisonous, and forms good crystals. It does not evolve any perceptible amount of hydrocyanic acid when treated with cold dilute sulphuric acid, but gives it off slowly on boiling.

Silver Cyanide, AgCN, is obtained as a white precipitate when a solution of potassium cyanide is added to one of silver nitrate. It is insoluble in acids, but dissolves readily in excess of the solution of KCN owing to the formation of a soluble double salt AgCN . KCN.

On this is based a method for the quantitative estimation of soluble cyanides; standard solution of silver nitrate is added to the solution of the cyanide until a permanent white precipitate just begins to form. When this occurs, one molecule of $AgNO_3$ has been added for every two molecules of the cyanide present:

$$AgNO_3 + 2KCN = AgCN \cdot KCN + KNO_3.$$

Double Cyanides.—In the double salt just referred to—$AgCN \cdot KCN$—we have an example of the marked tendency shown by various cyanides of different metals to combine to form double cyanides. In some of these double salts the combination is only loose and is readily broken, while their properties are not fundamentally different from those of simple cyanides. But in another important class of double cyanides the combination is so complete that the essential properties of the constituent salts entirely disappear in the double cyanide formed by their union.

Potassium Ferrocyanide, K_4FeCy_6, is such a double cyanide. Its formula may be regarded as showing it to be made up of $4KCy + FeCy_2$, four molecules of potassium cyanide with one of ferrous cyanide, but in reality neither the cyanogen group, nor the iron contained in the ferrocyanide can be detected by their ordinary reactions. The ferrocyanide is almost non-poisonous in comparison with the intensely poisonous nature of the soluble simple cyanides, and the iron in it is not precipitated by ammonium sulphide.

Potassium ferrocyanide is largely manufactured to serve as a starting-point for the preparation of Prussian blue and other cyanogen compounds. It is known commercially as yellow prussiate of potash, and is made by heating in shallow iron pans a mixture of charred nitrogenous refuse (horn, skin, etc.) with potash and iron filings. The ferrocyanide is extracted with water from the fused residue and purified by recrystallisation. It forms large tabular crystals of an amber colour. It dissolves easily in water, and the solution gives characteristic precipitates with solutions of several metallic salts, *e.g.* with copper sulphate solution a brown precipitate of copper ferrocyanide is obtained :

$$K_4FeCy_6 + 2CuSO_4 = 2K_2SO_4 + Cu_2FeCy_6.$$
Potassium ferrocyanide.   Copper ferrocyanide.

When a strong acid, (HCl), is added to the concentrated solution of potassium ferrocyanide, a white precipitate of *ferrocyanic acid*, H_4FeCy_6, is produced :

$$K_4FeCy_6 + 4HCl = 4KCl + H_4FeCy_6.$$

The complex radicle, $Fe(CN)_6$ or $FeCy_6$, which is present in ferrocyanic acid and the ferrocyanides, carries the name "ferrocyanogen."

Potassium Ferricyanide, K_3FeCy_6, is formed by oxidising a solution of the ferrocyanide by means of chlorine:

$$2K_4FeCy_6 + Cl_2 = 2KCl + 2K_3FeCy_6.$$

It may be regarded as built up from three molecules of KCN with one of ferric cyanide $FeCy_3$, but just as is the case with the ferrocyanide the properties of the compound are essentially different from those of the simple cyanides.

Potassium ferricyanides is the commercial "red prussiate of potash," and forms deep-red crystals.

Iron Salts and the Ferro- and Ferricyanides. — The reactions between solutions of iron salts and ferro- or ferricyanides are of importance in analytical chemistry, and for the thorough understanding of the composition of Prussian blue. They are best shown in a tabulated form:

Solution used.	Potassium Ferrocyanide. K_4FeCy_6.	Potassium Ferricyanide. K_3FeCy_6.
Ferrous salt	Light-blue pp. of ferrous ferrocyanide, which gradually darkens in the air	Dark blue pp. of ferrous ferricyanide; Turnbull's blue
Ferric salt	Dark blue pp. of ferric ferrocyanide; Prussian blue	No pp., but the solution becomes very dark green in colour

Prussian Blue, as indicated above, is chemically to be regarded as ferric ferrocyanide, $Fe_4(FeCy_6)_3$ or Fe_7Cy_{18}, and is formed when potassium ferrocyanide is added to a solution of a ferric salt. In actual practice it is made by adding the ferrocyanide to a somewhat oxidised solution of ferrous sulphate, and then completing the oxidation by means of air.

Cyanic Acid, HCNO. It has been mentioned that

potassium cyanide is a powerful reducing agent, and is used as such in analytical chemistry for the purpose of reducing the metals from their salts by fusion with sodium carbonate and the cyanide. In such reactions the potassium cyanide is converted by addition of oxygen into potassium cyanate:

$$KCN + O = KCNO.$$

The cyanate is more cheaply prepared by heating potassium ferrocyanide with an oxidising agent, such as MnO_2 or $K_2Cr_2O_7$.

Potassium Cyanate, $KCNO$, is a white solid, easily soluble in water. The solution gradually decomposes when kept. On addition of an acid free cyanic acid is not obtained, but only its products of decomposition with water—ammonia and carbon dioxide:

$$HCNO + H_2O = NH_3 + CO_2.$$

Ammonium Cyanate, NH_4CNO, is of special importance on account of its ready transformation into urea, $CO(NH_2)_2$, see p. 128. It is most easily obtained in solution by mixing strong solutions of potassium cyanate (prepared as above) and ammonium sulphate. The difficultly soluble potassium sulphate will separate out in part:

$$2KCNO + (NH_4)_2SO_4 = K_2SO_4 + 2NH_4CNO,$$

and the solution on evaporation yields urea along with some potassium sulphate.

Free Cyanic Acid, $HCNO$, has to be prepared indirectly. When solid urea is heated ammonia is at first evolved, but after a time ceases; if the residue is dissolved in potash solution, addition of an acid precipitates **Cyanuric Acid**, $H_3C_3N_3O_3$, which is produced according to the equation:

$$\underset{\text{Urea.}}{3CON_2H_4} = \underset{\text{Cyanuric acid.}}{H_3C_3N_3O_3} + 3NH_3.$$

If this cyanuric acid is collected and dried, and then heated in a retort, vapours of *cyanic acid*, $HCNO$, are evolved:

$$\underset{\text{Cyanuric acid.}}{H_3C_3N_3O_3} = \underset{\text{Cyanic acid.}}{3HCNO},$$

and can be condensed in a tube surrounded by a freezing mixture to a very volatile liquid, with a marked and acrid odour. Cyanic acid very readily undergoes polymerisation, forming either cyanuric acid or another polymer—cyamelide—whose molecular formula is uncertain.

QUESTIONS ON CHAPTER XX

1. Give three methods by which hydrocyanic acid can be prepared.
2. How would you proceed in order to obtain mercuric cyanide from potassium ferrocyanide?
3. What happens when you heat the following substances, (*a*) mercuric cyanide, (*b*) urea, (*c*) potassium ferrocyanide?
4. What is the composition of Prussian blue? Describe its manufacture.

INDEX

ACETALDEHYDE, 63
Acetamide, 89
Acetic anhydride, 79
Acetone, 66
Acetyl chloride, 79
Acetylene, 33
Acid, acetic, 68, 71
,, acrylic, 112
,, amido-acetic, 90
,, butyric, 74, 75
,, citric, 108
,, cyanic, 137
,, formic, 70
,, glycolic, 100
,, hydrocyanic, 132
,, lactic, 106
,, malic, 103
,, oxalic, 101
,, palmitic, 76
,, para-lactic, 107
,, propionic, 74
,, prussic, 132
,, stearic, 76
,, succinic, 102
,, tartaric, 103
,, uric, 129
Acrolein, 112, 116
Alcohol, allyl, 111
,, amyl, 54
,, butyl, 53
,, dihydric, 99
,, ethyl, 47, 51
,, methyl, 45
,, primary, secondary, and tertiary, 53
,, propyl, 52
,, trihydric, 115

Alcoholates, 51
Alcoholometry, 49
Alcohols, general characteristics of, 44
Allyl compounds, 111
Amides, 88
Amido-acetic acid, 90
Amines, 82
,, primary, secondary, and tertiary, 82, 84, 85
Argol, 103, 104
Arsines, 93
Asymmetric carbon atom, 108

BUTANE, 23
Butylene, 31

CACODYL compounds, 94
Cane-sugar, 123
Carbon, estimation of, 4
,, tetrachloride, 39
Carius's method of analysis, 9
Cellulose, 126
Chloral, 39, 65
Chlorhydrins, 118
Chloroform, 38
Couple, zinc-copper, 20, 95
Cyanides, 133
Cyanogen, 131

DEXTRIN, 126
Dextrose, 120
Dihydric alcohol, 99
Dulcitol, 120
Dynamite, 117

ETHANE, 22
Ether, 58

Ethereal salts, 55
Ethyl acetate, 56
,, alcohol, 47, 51
,, bromide, 41
,, chloride, 40
,, ether, 58
,, iodide, 42
,, mercaptan, 59
,, sulphide, 59
Ethylamines, 87
Ethylene, 28
,, bromide, 41

FEHLING'S solution, 62, 120, 122
Fermentation, 47
Formaldehyde, 61
Formulæ, empirical, 12
,, molecular, 13
Fusel oil, 47

GALACTOSE, 121
Garlic, oil of, 113
Glucose, 119
Glycerine, 115
Glycocoll, 90
Glycol, 99
Glyoxal, 100
Gun-cotton, 126

HALOGENS, estimation of, 9
Hofmann's method, 15
Homology, 2, 21
Hydrogen, estimation of, 4
Hydroxyl groups, determination of, 80

INVERT-SUGAR, 124
Iodoform, 39
Isomerism, 2, 23

KETONES, 65

LEAD ETHYL, 97
Levulose, 121

MALTOSE, 124
Mannitol, 120
Mercury methyl, 97
Methane, 19
Methyl alcohol, 45
,, chloride, 37

Methyl iodide, 39
Methylamine, 85
Methylated spirit, 49
Milk-sugar, 124
Mustard, oil of, 113

NITROGEN, estimation of, 6
Nitro-glycerine, 116

PARAFFIN, 27
Paraldehyde, 63
Pentane, 25
Petroleum, 26
Phosphines, 92
Phosphorus, estimation of, 10
Potassium ferrocyanide, 135
Proof spirit, 50
Propane, 23
Propylene, 30
Prussian blue, 136
Pyroxylin, 127

RAOULT'S method, 17, 119

SAPONIFICATION, 57
Saturated compounds, 28
Silicon, compounds of, 94
Soap, 76
Starch, 125
Sulphur, estimation of, 125

TARTAR, cream of, 104
,, emetic, 105
Tetrahedral carbon atom, theory of, 31, 35, 120
Tin ethyl, 98

UNSATURATED compounds, 28
Urea, 1, 128

VAN'T HOFF'S theory, 31, 120
Vapour density, 14
Victor Meyer's method, 14

WOOD, distillation of, 45

YEAST, 47

ZINC ETHYL, 96
,, methyl, 95